AF545064

Personal Protection and Security: A Practical Guide

Personal Protection and Security: A Practical Guide

Jay R. Dixon

Nelson-Hall nh Chicago

I affectionately dedicate this volume to my wife and best friend, Patricia.

Library of Congress Cataloging in Publication Data

Dixon, Jay R.
A practical guide to personal protection and security.

Includes index.
1. Crime prevention—United States. 2. Crime prevention—United States—Citizen participation. 3. Dwellings—United States—Security measures. I. Title.
HV7431.D59 1984 362.8'8 84-1043
ISBN 0-8304-1034-1

For information address Nelson-Hall Inc., Publishers, 111 North Canal Street, Chicago, Illinois 60606.

Manufactured in the United States of America

10 9 8 7 6 5 4 3 2 1

The paper in this book is pH neutral (acid-free).

Contents

1 Introduction

There is much crime in America, more than ever is reported, far more than ever is solved, far too much for the health of the Nation. Every American knows that. Every American is, in a sense, a victim of crime.

Written in 1967, that is the preamble to *The Challenge of Crime in a Free Society,* a report by President Lyndon Johnson's Commission on Law Enforcement and the Administration of Justice, a body charged with finding ways to reduce crime. The crime rate has not improved since that report was written. It has gone from bad to worse. It is hard to find an American who has not personally been the victim of some crime. The attorney general of the United States recently told Congress that, last year, about one of three households in the country was victimized by some form of serious crime. If the trend continues, he said, within a few years every family in America will personally experience the outrage of violent crime.

Articles about crime appear daily in newspapers. On television and radio we see and hear about the increase in violence. Once a week in the mail, or stuck to the front door, we receive advertising from alarm companies, guard dog agencies, and mace training classes. They cite statistics, valid statistics like burglars are successful in 80 percent of their attempts and each of us has one chance in six of being a burglary victim within the next year.

Crime has become such a routine event that many of us don't even bother to report it. Federal agencies estimate that two-thirds of all violent crimes are never reported. One reason they are not reported is that a good number of people believe nothing will, or can, be done about crime. They feel helpless.

We have depended upon our governmental agencies to solve the crime problem for us; we, and they, have failed miserably. In recent polls, the fear of crime placed near the top of the list of concerns of a majority of Americans.

One of the reasons we have failed is because we have expected others to do the job for us. Those others have been the police, prosecuting attorneys, courts, prisons, probation and parole officers, lawyers, and legislators. We delegated responsibility for crime to the legal system. Crime was their problem to solve, so long as it didn't touch us directly. But now it is touching us. We don't like it, but we don't know what to do about it.

Perhaps we failed because we focused only on the offender. It is he or she who is responsible. It is he or she who, if punished, won't do it again, who, if "corrected," will become a useful citizen. Concentrating on offenders has not worked and is not likely to work, because there are now more offenders than we can cope with. The more we put away, the more appear. They seem to multiply like rabbits. One of every eighty-three adults in this country was in jail, on probation, or on parole in 1978, and the number has grown since then.

We have now reached a point where, if we expect to be protected, we must do it ourselves. That is not to say that we must return to the days of the lynch mobs; we have matured as a society beyond that phase of our history. But we can no longer depend upon government to protect us. Even if their techniques worked, and they haven't, government no longer has the resources. The results of a recent Northwestern University study, for example, show that the number of police officers in this country fell from 3.32 for each reported violent crime in 1948, to 0.5 in 1978. We must protect ourselves.

The reason for this book is to tell you about ways in which you can protect yourself. You can do it without a large capital outlay and without turning your home into a Fort Knox or developing a security force equivalent to that needed to protect a nuclear power plant.

Other books have been written on the subject. There are books from the criminal's point of view, the police officer's point of view, the psychologist's point of view, and the sociologist's point of view. Some are theoretical, some philosophical, some pollyannaish. This one is practical.

I've avoided time worn advice which, while well intended, is unlikely to be followed. You will find in succeeding pages a prescription for a course of action that can reduce your chances of becoming a crime victim and increase your likelihood of surviving an attack. It is a collection of ideas that form a strategy for the prevention of crime against you.

2 A Moat for Your Castle

Some of us fondly remember when it was unnecessary to lock the doors of our homes, let alone the windows; a time when churches were sanctuaries and left their doors open day and night; when someone who rang the doorbell asking for help was probably honestly seeking it; when crime was something we read about in novels or newspapers and occasionally happened to someone else far, far away. Those *were* the days.

Today, your home requires a level of protection that will establish a barrier between you and your possessions and the outside world. That barrier should deter and delay a would-be criminal and provide you and the police time to react. Alarm systems are discussed in another chapter. Here we'll devote our attention to deciding where you should live, to physical barriers, to lighting, and to other general tips. This advice will apply whether you live in a house, an apartment or condominium, or a travel trailer.

Although economics and other factors play a part in deciding where you live, many people have the option to continue to live in their present neighborhoods or to move elsewhere. I promised to try to give practical advice, and moving may not be practical for you. But the point I want to make is that you are in control of your life. Where you decide to live is an important factor in your life, and moving or not moving is often a question of setting the proper priorities. If crime in your present neighborhood is rampant, moving to a different neighborhood may be more practical than attempting to establish your own crime-free oasis.

If you are making a move, for whatever reason, consider how you are going to decide where to live from the point of view of self-protection. You can't judge a neighborhood just by its appearance or by intuition. On a larger geographical basis, for example, try this short quiz. In which of our fifty states is the murder rate highest—California, New York, Michigan, or Illinois? Answer: none of the above. At this writing, the highest per capita murder rate is in Nevada, followed by Texas, Louisiana, and then California. Illinois and Michigan aren't among the top ten states.

You might look at your odds of survival in certain cities. In the United States, someone is murdered every twenty-four minutes, robbed every fifty-five seconds, and a victim of aggravated assault every forty-nine seconds. Your chances of becoming a homicide victim are highest in Miami, St. Louis, Newark, or Atlanta, and lowest in Minneapolis, El Paso, Tucson, Omaha, and Indianapolis. If it's robberies you're wary of, stay away from Newark, Miami, Boston, and Washington D.C., and you're safest in Austin, El Paso, San Antonio, or Charlotte. Aggravated assaults are highest in (you've heard these names before) Miami, Atlanta, Newark, and St. Louis, but lowest in Honolulu, Omaha, Austin, and Toledo.

About one in every thousand persons over the age of twelve is raped each year in this country, seven of every thousand are robbed, eighty-four of every thousand residences are burglarized, and seventeen of every thousand households lose a motor vehicle to a thief. A burglary occurs every ten seconds; burglary is the most frequent crime in America. Somewhere between two and four million burglaries occur here annually. Since not everyone reports them, statistics are incomplete, but the total annual take from these crimes is estimated to be billions of dollars. In an average burglary, the loss is about $900.

Your chances of being burglarized are higher if you rent than if you own your home; if you live in the city rather than the country; if you live next to an open public space like a park, playground, parking lot or alley. According to a Canadian study, most criminals are lazy. They can afford to be, because the design of most communities makes crime easy. Most thefts occur along popularly traveled routes where strangers are commonplace. Crime is lower on cul-de-sacs where the houses face one another and intruders are more easily detected.

Crime statistics are available in your local library. Many large cities collect what is called crime census tract data, which will tell you the crime rates in the geographical areas in which you have an interest. Other cities keep track of crime by police districts or precincts. Few people bother to find out what the crime rates are in the cities or states in which they decide to live, let alone their neighborhoods. Before you move to an area, take time to check the statistics and trends. Don't just look at the crime rate in your neighborhood; see how that rate stacks up against other neighborhoods. There are no communities that are free of crime. What you are looking for is the best location in the area in which you want to live—a location where the crime rate is relatively low.

Once you've selected the part of town in which you want to live, take a hard look at the physical characteristics of your home. Is it hidden from the street? That provides a measure of privacy, not just for you, but for the prowler, burglar, or rapist as well. If you must live in seclusion, you'll need to take extra precautions to raise your home protection to an acceptable level. It's better if your home and its entrances can be seen by passersby, including the police and the neighbors. If there is shrubbery that inhibits that view and provides hiding places for intruders, trim it or remove it.

If there are features of your home or landscaping that make it easy for someone to enter, such as a tree or trellis that provides a route to a second story window, trim the tree or remove the trellis. And don't contribute to the ease of a second-story job by leaving your ladder where it is accessible to the burglar. Although intruders can climb over fences or gates, it's a good idea to lock the inside of a gate with a hasp and padlock to make entry more difficult. If you have a swimming pool, it should be surrounded by a fence and the gate should be kept locked.

Make sure your house number is prominently displayed so that emergency personnel can see it from the street. It's well to have the number in several places; near the front door, the garage door, and the street. Painted on the curb may be a good location, but remember that parked cars may obscure it. An illuminated number is helpful at night. If there is an alley behind your home, the number should be there also. In cities where there is a police helicopter patrol, the police may recommend having the number on your roof, as well. There are companies specializing in that kind of painting, but make sure that having the numbers on the roof will be helpful to the police before you spend your money.

If you live in a multiple unit dwelling, your chances of being burglarized are less if access to your building is controlled by something more than a key. That means an access control system of some type, closed circuit television, security personnel, or some combination of the three. Don't give keys to more people than you are required to under your lease or rental agreement. While your neighbors may live close to you, don't develop a false sense of security.

Because you live on an upper floor does not mean that you are immune from burglars. Take a look at your outside patio or balcony to see whether someone could get to it from an adjoining balcony or from one above or below it. If they could, you are relying upon your neighbor's protective barriers unless you also protect your space.

Whatever the type of residence in which you live, exterior doors should be solid barriers. They should be either of hollow core steel construction or made of solid wood at least two inches thick. Doors with glass panels don't provide protection. It's simple for an intruder to break the glass, reach inside, and unlock the door. Glass panels alongside the door pose the same problem. Replace glass paneled doors with solid doors, or replace the glass with an unbreakable acrylic. If you must have a window in the door, cover it from the inside so that persons standing on your porch cannot see you when you approach the door. On the other hand, since you should always know who is on the other side of your door before you open it, install a wide angle peephole in each solid exterior door. A peephole is better than a window because you can see out without being seen.

An intercom works for asking the identity of a visitor, but it does give away your presence in the home. And, unless you recognize the voice, you won't be able to identify the visitor.

Examine the outside of your doors closely. When the doors are closed, the hinges should not be visible from the outside, nor should the hinge pins (the bolt-like objects that hold the two halves of the hinges together). If they are, the burglar can unscrew the hinges or slip the hinge pins out and doesn't need to bother to break in. He can just lift the door out of its frame. There are two ways to correct that. The best way is to properly install the hinges on the inside of the door. If that is inconvenient or expensive, a stopgap measure is to remove a screw from each side of the hinge. Remove one screw from the side of the hinge fastened to the door, then remove the screw directly opposite it from the side of the hinge fastened to the door frame. Insert a short piece of metal rod (a nail, screw or bolt may work) in the hole in the door frame, so that it sticks out about half an inch. When the door is closed, that rod will fit into the hole in the door, and will keep the door from being lifted out of its frame.

Sliding glass doors, common in newer homes and apartments, present several problems. Latches on most can easily be forced and the door slid open. You can install an inexpensive antislide lock, or you can put a piece of broomstick, metal rod, or doweling snugly in the lower track. With that in place, the door will not slide open.

Many glass doors slide on an outside track, rather than inside. That makes it easier for an intruder to lift the door out of the frame, even while it is locked, and a piece of broomstick stuck in the track on the outside won't help you. To deter a lift-out, install a lock at the bottom of the door that has a hook that holds the door to the threshold or frame. Or use a lock that mounts on the edge of the sliding door panel and attaches to the edge of the stationary panel. When locked in this manner, the door cannot be slid open or lifted out. You can also drill a hole through the side of the top track into the frame of the sliding panel and insert a bolt or pin into the hole when the door is closed. And, you can screw several self-tapping screws into the top of the top track just above the door. The screws take up clearance and prevent the door from being lifted out. If you need to remove the door, you simply remove the screws when the door is open.

Treat doors between a garage or basement and the house as you would any other exterior door. Never leave your garage door open, and lock it when it is closed. Because of the size and construction of the doors, good strong locks for garage doors are hard to find, but you can use hasps and high quality padlocks on both sides of the inside of the door to lock the door to the frame.

If you use an automatic garage door opener, check to make sure it does not open for radio transmitters from vehicles on the street, or airplanes flying overhead, or when a neighbor's garage door opener is used. You can purchase a multifrequency garage door opener that will allow you to change your frequency if you experience such problems. Check your door to see if it can be pulled up from the outside. If it can, adjust the mechanism until there is no slack when the door is closed. That will stop a thief from crawling through the space under the door. If a thief does get into your garage, he or she can lie in wait for you or can spend whatever time is necessary, using your tools, to gain entry to your home without being observed. So do your best to protect the garage as well as the rest of the house.

If your home has large windows, cover them with drapes, shutters, or blinds, but leave these coverings open when away from home and when privacy isn't necessary. If a stranger is wandering through your home, you will want the neighbors or police to be able to see him.

Windows can be used to enter your home just as easily as doors. If windows are large enough to crawl through, a thief can either break the glass and crawl in, or break or cut the glass, open the frame, and crawl in. If you don't have a need to open double-hung windows (those which slide up and down), nail them shut. You can break the glass from the inside if you need to use a window as a means of escape from the home. If you occasionally open a window, install screws in the window casing that will allow the window to open, but not far enough for someone to crawl through. An alternative to nailing a window shut is to drill a hole through the top of the lower window frame into the bottom of the upper window frame and insert a pin. That will lock the window shut but still allow you to open it by removing the pin.

If you have sliding windows of aluminum or steel, protect them from being lifted out from the outside by installing a self-tapping metal screw in the top track, as you did with your sliding glass door. The window will still slide open when you want it to, but it cannot be lifted out unless the screw is removed, and the screw can be removed only from the inside. A piece of broomstick, dowel, or metal rod, placed in the track next to the sliding window, will prevent someone outside from forcing the window open.

Casement windows are those which crank open and closed. If the latch works properly and there's not excessive play in the crank, they are reasonably secure.

If you have louvered glass windows, it's best to replace them with solid glass. The glass sections of a louvered window can be removed easily from the outside. While you can try gluing the panes in with an epoxy glue, my advice is to replace the windows. In older homes, window glass is usually held in the window frame with putty, which will dry out over time. When the putty is dry, an intruder can scrape or chip it out and lift out the pane of glass. If you have this type of window, have it reputtied periodically, or do it yourself with inexpensive materials from your local hardware store.

There is no real protection against the breaking of glass in a window or door, but you can replace glass with an unbreakable acrylic material. While it is more expensive than glass, it is clear, and properly maintained, should last for years.

If there are windows in your garage, protect them as you would protect other windows in your home. Doors and windows should be closed and locked whenever there is no good reason to have them open. Never leave home with doors or windows open, not even for "just a minute." In half the burglaries in this country, the burglar enters through an unlocked door or window.

When you move into a new home, change all exterior locks. That doesn't mean that you have to throw away all the existing hardware; a locksmith can put new cylinders into existing locks. Use locks or cylinders that have keys that cannot be duplicated at your local key kiosk. Don't be misled by keys that are stamped "do not duplicate." That message is psychological at best. It does not constrain anyone from cutting a new key from the old one. You should buy a brand of lock, usually sold only by locksmiths, whose manufacturer controls the key blanks. While there is no absolute protection against copying, you will put the would-be thief to much greater difficulty by buying and installing that type of lock.

All exterior doors should be equipped with deadbolt locks with at least a one-inch throw. The throw is the part of the bolt that extends into the door frame or jamb when you lock the door. Make sure the frame or jamb is as solid as the door, or your lock will be useless. Most deadbolts can be locked from the outside only with a key. That means you'll have to stop on the way out to lock the door, just as you do on the way in to unlock it. The inconvenience will be well worth the trouble, since most other locks can be easily defeated.

There are several kinds of deadbolt locks. For safety's sake in an emergency, it's well to have one that works in combination with the doorknob. While inside, if you need to leave the house in a hurry, as when there's a fire, turning the inside knob retracts the deadbolt as well as the door latch and leaves the door unlocked.

A more common deadbolt lock is the mortise lock. It is separate from the doorknob lockset, usually above it, and is opened by turning the deadbolt knob on the inside of the door. In an emergency, if you have this kind of lock, you will have to unlock both the deadbolt and doorknob locks in order to open the door.

Another type of deadbolt is the rim lock. Instead of being installed in the door, it is installed on the inside of the door and has a bolt that slides into a piece of metal attached to the door frame. These are usually used as auxiliary locks where the installation of another type would be too expensive or inconvenient.

Commonly used inexpensive doorknob latch sets, with an outside keyway in the center of the knob and a turn button on the inside knob, and no deadbolt, are easily compromised. If the latch cannot be manipulated from the outside with a piece of plastic, such as a credit card, the knob can be twisted off with a wrench or water pump pliers. The best locks have a deadbolt of hardened steel, making it difficult to saw through, and a free-spinning exterior knob that cannot be twisted off.

Some people recommend deadbolt locks that are keyed on both sides. They would prevent someone from breaking glass in a door, for example, and reaching in to unlock the door. While there are places for such locks, the exit from your home is not one of them. Whatever protective devices you install, you want to be able to get out of the house in a hurry in the event of fire or other emergency. You don't want to be searching for the key or the keyhole, in the dark, in order to escape. And if you leave the key in the lock on the inside for such occasions, what good is the lock?

Some suggest you have a deadbolt lock on your bedroom door, to further protect you if an intruder successfully enters your home. If you live in a rooming or boarding house, a deadbolt makes sense because that door is really your exterior door. It probably doesn't make sense if you live in a single family home. If the intruder has been able to enter the house, the lock may not stop him from kicking in an inner door, particularly since he will be unobserved as he does so. Most inner doors have hollow cores and thin wood, and they are easily kicked open or broken through. On the other hand, the lock may delay your getting out of the house in an emergency and could impede fire or other personnel who may try to reach you.

Don't rely upon chain locks for protection. A good hard push against the door will usually rip the chain mechanism right out of the frame. While you may have slowed an intruder down a bit and caused him to make some noise, it is likely that he'll still get inside.

Security grates and grills and bars on windows are prevalent these days, particularly in high crime areas. While they can be an effective security measure, they can also create a death trap. In a fire, earthquake, or other emergency, you may need to use your windows to exit your home. If you use these devices, make sure they have quick release mechanisms and that those mechanisms work. The release mechanisms must be located so that a burglar can't reach through the grill or grate, break the glass, and operate the mechanism. Bars, grills, and grates will increase the time it takes emergency personnel to get to you when time counts the most. Don't trade protection from a burglar for the risk of being unable to escape a fire.

In your protective plan, don't overlook other means of entry into your home. A cellar door, a patio door, a skylight, and, in older homes, a milk or coal chute all need to be protected. If you use a window-mounted air conditioner, it can be lifted out unless you secure it to the window frame. Also, although it's unlikely that anyone will be able to enter your home through your outside fuse box, keep a padlock on it, so that pranksters and others who might want to leave you in the dark can't tamper with your fuses or circuit breakers.

Should you have a watchdog to protect yourself or your belongings? I know a fellow who has a pet cougar; that would keep me from sneaking into his house! Parrots and macaws are being promoted as "watchbirds" in New York. Pet shop owners claim that trained parrots and macaws can bark, growl, bite, and make just as much noise as a dog when someone tries to break in. A four-foot cobra was used recently to guard a $1.8 million sapphire in London. Unfortunately it so charmed onlookers that thieves were able to steal jewels from nearby displays while attention was focused on the cobra.

Noise and light are two of your best weapons against an intruder, and an animal may provide you with a noisy early warning system. However, unless your dog is trained to be a fighter, don't depend upon it to do more than bark. Most pets can be distracted or are just too friendly to keep a burglar at bay, but a properly trained dog can be effective in discouraging intruders.

The physical characteristics of a dog contribute greatly to its effectiveness. If it looks ferocious, it may be as effective as if it is ferocious. Large dogs—Dobermans, German Shepherds, or Great Danes—have both a threatening appearance and a natural tendency to protect their keepers.

Those who want a dog to have more than appearance or natural instincts going for it should have the dog formally trained. A dog can be taught to bark, snarl, or molest an intruder, but proper training requires a substantial investment in patience, time, and money. A dog that is poorly trained or has been used in military or police functions may be a liability. It may protect you or it may attack friends, neighbors, and even you.

Proper training involves teaching the owners as well as the dog. Avoid training courses that offer to take your dog for a couple of weeks and return it to you fully trained. That kind of training program may return a dog that will turn on you and your loved ones as easily as it will turn on an intruder. A responsible trainer will test your dog and candidly tell you whether it can be trained as a protector.

If you're looking for a protector you won't have to feed, scientists say the first security robots should be on the market within a year or two. They will be able to move around on wheels, to ''hear'' people through walls, and to detect sounds like the breaking of glass. Robots may be able to do their job better than human guards. For one thing, they won't fall asleep on the job. Security robots will have microwave and infrared sensors and sonar range finders to help them avoid walls and other obstacles and microcomputers to process the information the robot receives from its sensors. Despite all that, robots won't be terribly intelligent. Since they won't be able to tell the good guys from the bad, they won't be allowed to associate with people. And if you try to steal a robot, it will scream through a 120-decible siren that is loud enough to wake all but the dead.

Making your home *look* occupied is often an effective deterrent. Because burglars don't like occupied homes, half of all burglaries are committed during the day when the home is empty. So, if you must be gone (and few people stay home all the time), do things that will disguise your absence. Never leave a note on the door telling where you have gone or when you will return. Don't leave notes for mail or delivery people. Turn the bell on your telephone down, so its ringing unanswered can't be heard from the outside. You can leave toys, tricycles, and other items outside your home to make it look occupied, but remember—someone might steal them.

Years ago I was able to say that keeping an outside light burning all night cost less than one cent. I can't say that with a straight face these days, but keep the light lit at night anyway. Lighting is still one of the least expensive crime deterrents you can buy. Consider replacing your outdoor fixtures with those that will accommodate a floodlamp that can illuminate a large area.

You can equip outdoor and indoor lights with sensors that are activated by darkness, or you can use a timer that will turn them on and off for you. Both save you the trouble of flipping the switch, and they operate the lights in darkness whether or not you are at home. You will probably also save money since the sensor or timer will turn the lights off when they are not needed.

Don't leave just a porch light on when you are gone at night, and don't leave outside lights on during the day. These are giveaways that the house is unoccupied. Leaving a television or radio on in the home may give the impression that someone is at home, and some folks have even purchased police scanners that they leave on all the time. There is nothing like the sound of police calls to frighten off a burglar.

Don't hide keys outside the house. Any place you are apt to hide a key is likely to be the same place a thief will look for it. Under the mat, in the downspout, or over the door ledge are well-known hiding places both to you and to intruders. Get used to locking your doors when you leave and taking your keys with you, and give keys to those who need them so you don't need to hide one somewhere. But, don't give a key to every Tom, Dick, and Jane. Few people need access to your home when you are not there. If you're worried about being locked out of your house, give a key to a friend or neighbor. If you or someone to whom you entrust a key loses it, be sure to have your locks changed. Normally, that only means modifying the cylinder to accept a new key, not replacing the whole lock.

Speaking of hiding places, the places the average person selects to hide valuables are usually the places a burglar will look for them. He'll look first in the top dresser drawer, the desk drawer, the night table, and then in the shoe boxes in the bedroom closet. So, if you must hide something in your home, at least be a little creative in your selection of hiding places. For example, if you need to keep cash at home, put it in a plastic bag or envelope and place it inside a commercially packaged frozen food container in your freezer. Of course, you'll have to remember that it's there before you stick the package in the microwave oven.

When you are away from home for any length of time, someone should know where you are and when you plan to return. Tell trusted neighbors and the police when you leave for an extended period. Some police departments will conduct a regular patrol of your home while you are away. Cutting off deliveries will inform people you may not know that you will be gone. It's better to arrange to have someone house or apartment sit, or to arrange for a neighbor or friend to take in the milk, mail, and the newspapers, and to water and mow the lawn. If you live in snow country, arrange to have walkways shoveled while you are away. Leave lights on in the home at night, or have someone turn them on and off periodically. Don't leave the same lights on all the time—that's a giveaway. And leave your draperies open. If someone does get inside, it will be more difficult for them to move around unseen.

If you live in a multiple-story building, check out the fire stairs. Make sure they are locked from the stair side, but that you can enter them from the apartment side. You will be able to get out if there is a fire, but an intruder can't use them to go from floor to floor. Particularly if you are female, don't list your full name on the mailbox, intercom, or doorbell. Those are good places to use your first initials and your last name.

Finally, don't keep things in your home that belong in a bank vault or in the granite vaults in the mountains of Utah. You should use and enjoy your possessions, but if they include the Hope diamond, make some banker happy and rent a safe-deposit box. Better yet, you can make him even happier by depositing enough money in his bank that he'll loan you a safe-deposit box for nothing. Inventory the valuable possessions you keep at home, keep a list of serial numbers of those that are so numbered, photograph your collectibles, and keep receipts and appraisals indicating their value.

Many people advocate engraving some personalized form of identification, such as your social security or driver's license number, on valuables, and some police departments and community organizations will loan you the engraving tools. If scratching your name or number on your valuables doesn't appeal to you, there is a security marker the size of a pen which contains a special "ink" that can be seen only under blacklight. Besides the obvious aesthetic advantage, the marker will write on cloth and glass as well as other surfaces. Some community organizations provide stickers to put on your doors and windows that warn intruders that your belongings are marked. You certainly should have some way of identifying your property. Your pocket camera will look just like anyone else's without a serial number or identifying mark.

For a small fee, Operation Identification, a program of the International Association of Chiefs of Police, offers property owners an engraving stylus and registration of valuables plus, if there is a burglary, a $500 reward for the apprehension of the burglar and $100 toward the member's household insurance deductible. If there's a fire, they offer a $2,500 loan interest free for thirty days. Each of the half million subscribers to the service has a unique identification number, which police officers can verify through a twenty-four hour toll free number.

There are services that will come to your home and describe your belongings on video tape, in brilliant color and resonant sound. Put the tape, and the rest of your records, in your safe-deposit box.

Should you have a safe in your home? There are two types of widely marketed home safes. The least expensive and most common is a fire resistant safe constructed of an insulating material and sheet metal. It will protect paper, but not items like magnetic tape or film, against fire. Look closely at the fire rating and at the period of time and the temperature against which the safe you are considering will protect papers. While such a safe may discourage an amateur burglar, they are not burglar-proof. Many are so small and light that a thief can simply take the safe with him.

Burglar resistant safes are made of solid steel, which means they may not protect against fire, since steel is a good conductor of heat. These safes may be installed in a wall or floor, but a floor is preferable, since most walls are not deep or strong enough to house a safe. Floor safes should be encased on all four sides in concrete. You'll find such a safe hard to take with you if you move.

If you decide to buy a safe, buy one of sufficient quality, size, and weight that it can't be easily opened or carted off. Industrial safes can provide both burglar protection and fire protection, but a safe-deposit box, while less convenient, will usually be a better deal.

Make sure you are adequately insured. Most homeowner's policies limit coverage of antiques and jewelry to a small dollar amount in relation to the rest of the coverage. Examine your policy carefully, and buy extra coverage if necessary. Don't keep large amounts of cash in your home. You'll have to determine what "large" means to you. It depends upon your resources. Cash belongs in the bank. Most insurance policies limit reimbursement for missing or stolen cash.

Given enough time and money, anyone can make their home almost completely secure from intruders. You would quickly find, however, that you would not, or could not, live in a perfectly secure home, even if you could afford one. A completely secure building is one which doesn't allow anyone in or anyone out. Since that's not how most of us want to live, only through taking reasonable precautions can we reduce our odds of becoming a victim. A high percentage of burglars are amateurs; over 80 percent of them are under the age of twenty-five. The average burglar spends less than three minutes in a home. By following the practical ideas just presented, you can outwit an amateur and maybe even a professional.

3 Our Neighbor's Keepers

In the arena of crime prevention and protection, we are all our brother's keepers. At least we should be, for our own benefit. In the *Introduction to Law Enforcement and Criminal Justice,* authors German, Day, and Gallati wrote, "In early tribal and clan life, the people were the police." It was in those tribes and clans that the concept of "kin police" developed, where the family, tribe, or clan assumed some or all responsibility for seeing that justice was done.

In the early history of civilization, there are many references to the responsibility of the people to protect themselves. When asked to name the essential ingredient of the ideal community, the Athenian legislator Solon replied, "When those who have not been injured become as indignant as those who have." He clearly implied shared social responsibility. Augustus of Rome required the residents of his city to assist his Praetorian guards in maintaining order.

Our system of law enforcement and justice is largely derived from the English system. Englishmen formed themselves into groups of ten families called tithings and shared the responsibility for maintaining the peace and protecting their community. Each individual in a tithing was responsible for the behavior of his neighbor.

The hue and cry, perhaps the forerunner to today's concept of citizen's arrest, required that every able-bodied man lay down his tools and join in the chase and apprehension of an offender. The watch and ward, in place in London in 1285, required that each household take a turn standing guard at a city gate. In our own country, a form of protection similar to the watch and ward has sometimes been practiced; in some towns, all able-bodied males over the age of sixteen were required to volunteer their services as watchmen. The first daytime, paid police in the United States appeared in 1833 in Philadelphia.

The serious student of law enforcement and criminal justice can explore in depth elsewhere the origins of our police system and of an individual citizen's responsibilities for the protection of others. My point in including this brief historical backdrop is to emphasize that, while we think of protection as a police or governmental responsibility, it has always been a personal responsibility as well.

The fact that, under certain conditions, a private person may legally make an arrest for certain offenses is a reflection of the responsibility of private persons for enforcing the law. In many jurisdictions a private citizen is required by law to go to the aid of a police officer when that aid is requested.

For political and economic reasons, we are faced with the reality of having to protect ourselves. One of the best preventive strategies we can develop is to know our neighbors and to let them know us and our habits. Even if you don't and won't socialize with your neighbors, let them watch the outside of your home and do the same for them. Know their names, addresses, and telephone numbers, including their work or emergency telephone numbers. That will help you give the police accurate information if you must call them because something goes wrong at a neighbor's home. Be sure to keep emergency telephone numbers of the police and fire departments near each of your telephones for that purpose. Exchange work and vacation schedules with your neighbors so that they know when you should and should not be home, and so you'll know the same about them.

While it may be called something different in your community, one of the most effective crime prevention efforts is the neighborhood watch program. It involves neighbors getting to know each other, working together to prevent and report crime, being trained to recognize and report suspicious activities, cooperating to implement crime prevention techniques, and generally looking out for one another. It's really a revival of the watch and ward program mentioned earlier, so it's not a new concept. It can be very effective and should not be confused with nosiness or vigilante efforts.

Vigilance is a key word in the concept, however. Since no society can ever have enough police officers to watch all private property, neighbors are recruited to do it. Anybody can participate, no matter their age or station in life. In some communities, police crime prevention officers may help you to organize the neighborhood. Some communities have crime prevention councils, and some governmental agencies have social workers or other civilian employees to assist you.

You should usually start by contacting a local law enforcement agency, and then have a neighborhood organizational meeting. Someone, of course, has to go door to door to discuss the crime situation in the neighborhood, explain the value of the program, schedule the first meeting, and get your neighbors interested enough to attend. That someone may turn out to be you.

At the first meeting someone should be selected to be the neighborhood spokesperson—the individual who serves as liaison between the group and law enforcement agencies, keeps the directory up to date, explains the program to new neighbors, and arranges periodic meetings of the group. Police departments, offices of attorneys general, insurance companies, and others will often provide crime prevention materials and even stickers to put on doors and windows that let passersby know that the community is organized against crime.

Someone needs to map the neighborhood to provide a diagram of the streets, the houses upon them, the house numbers, the names of the residents, and their telephone numbers. This becomes a neighborhood directory that will enable residents to give correct information when reporting suspicious activity.

Nothing in the program should be designed to encourage individuals to take personal risks to prevent crime or to shift the responsibility for apprehending criminals from the police to private citizens. Catching crooks is the business of the police, but preventing crime is everybody's business.

It's important for everyone in a neighborhood to know one another. That provides a feeling of group solidarity, security, and shared responsibility. Although there may not be 100 percent involvement, neighbors who know one another are more likely to go to one another's aid and to accept responsibility for their neighbor's safety. Beyond direct intervention, neighbors need to be willing to pick up the telephone and call the police if something suspicious is occurring.

The neighborhood watch program encourages neighbors to gather and share information about one another. Besides knowing your neighbors' names, addresses, and telephone numbers, you may also want to keep track of the number, ages, and identities of family members; work and school hours; numbers and descriptions of family vehicles and pets; normal delivery times; and planned vacations or visitors.

Some neighborhood groups form citizen patrols—small groups of residents who walk the neighborhood streets, especially at night. They report suspicious activities, escort senior citizens, and make noise to ward off would-be offenders. Some groups use mobile patrols and citizens band radios to communicate with one another and the police. While each community group will develop the programs it needs and in which its members are willing to participate, it is important to remember that *involvement* is the key to the success of the neighborhood watch program, not the sophistication or number of its activities.

Neighborhood watch programs are sometimes organized by the victim or victims of a crime. My first involvement was in the 1960s, when a Seattle couple was robbed at gunpoint, and she was raped. The husband, working with a local insurance company, the police, and neighbors, began a very effective crime prevention effort there. The outraged and humiliated couple turned a horrifying experience into one which ultimately benefited the entire community.

Another type of neighborhood program is called a block parent program. It's sometimes part of a neighborhood watch program, and I'll tell you more about it in chapter 13, entitled Dangerous Stranger.

To reduce crime and assure a quieter neighborhood, citizens in some cities close off some streets to all but pedestrian traffic. To some extent this is reminiscent of the ancient practice of walling off a city, but most of these efforts have done little more than limit vehicular traffic to certain streets. Because everyone walks, residents mingle with their neighbors, keep an eye on each other, and watch each other's property. Some private streets have been walled or fenced off and some are simply blocked with barricades. On others, traffic is controlled by restrictive signs. Crime rates decrease in neighborhoods where this has been done, perhaps because a thief has less opportunity to use a vehicle to make a fast getaway after a crime.

In at least one community—Miami Beach, Florida—remote television cameras are used to observe main traffic arteries. One hundred camera housings have been installed, and twenty-one cameras are in place. Criminals cannot tell which of the housings have cameras and which do not, and the police change them around from time to time. All activity on the streets actually observed by cameras is video recorded, but the real impact of the program is expected to be crime deterrence, not the gathering of evidence.

A California program which goes beyond the neighborhood is called WeTiP.[1] Volunteers operate telephone lines and anonymous

1. Copyright 1981 WeTiP, Inc.

callers can offer information in exchange for the possibility of a reward if the information leads to the arrest and conviction of a criminal. Callers are assigned code names and numbers, and the rewards are paid in cash. The reward fund comes from donations from service clubs, community groups, businesses, and individuals. Other states have similar programs under different names.

By collectively participating with your neighbors and fellow citizens in crime prevention activities, you can help achieve results beyond those that you could accomplish individually. Where neighborhood programs are operational, crime rates have sometimes been reduced 40 percent and more. When that volunteer comes knocking on your door, get involved. Better yet, why not become that volunteer yourself!

4 Should You Be Alarmed?

Today there is a home alarm system for every pocketbook. You can spend from $10 to thousands. But, before you run off to buy one, there are some questions you should ask yourself. What do you want to protect? Do you want a system that will frighten off a would-be thief or one that will help catch him or her? Do you have enough valuables in your home to justify an elaborate system? Is it your safety you're concerned about or protecting your property when you're not home?

The size of your budget, your specific protection needs, and whether you have small children or pets will determine the kind of alarm system you buy, install, or have installed. If you simply want the appearance of an alarm system, you can put alarm company stickers on your doors and windows and hope no one calls your bluff, but I don't recommend bluffing. If you are interested in a real alarm system, read on.

An intrusion system protects the perimeter of your home. Such systems are normally "hard-wired." Small devices known as contacts are placed on doors and windows to sense when the doors or windows are opened and then wired to an alarm control box somewhere in your home. The alarm control box will activate a bell or siren located either inside or outside your home and may also send a signal through a direct wire to a security service central alarm station or to the police. Except in rural or very small communities, most police departments no longer monitor alarms. It's not that they don't want to know when you have an intruder, but, with the press of other business, they've gotten out of the alarm business. They may or may not be able to recommend an alarm monitoring company to you.

Some alarm systems are connected to a central alarm station through a telephone line for which one pays a monthly fee. A system may be hooked to a loop telephone line which connects a number of alarm subscribers, or the alarm may use a regular telephone line to transmit a recorded message to the police. The latter method is known as a dialer and is the least desirable of the three, since simply cutting your telephone line outside your home will prevent your alarm signal from reaching the central alarm station. You'll have no way of knowing it's cut unless you try to use the telephone and find the line dead. And, even when you do know about it, you may still be out of luck. (There are some dialer systems which will activate a local alarm if the line is tampered with.) A loop is a somewhat more desirable system, but if there is trouble anywhere on the loop, it may take the alarm or telephone company some time to find the trouble and fix it, and your alarm may not work in the interim. A direct telephone line between your home alarm system and the central station is the most desirable but also the most expensive method. In areas where the physical terrain permits, home alarm signals may be sent to a central alarm station via radio transmitter. If available, that method is often more trustworthy and less expensive than telephone lines.

Door and window contacts can sense when doors or windows are opened and begin a chain of events that ultimately summons help. However, the contacts will not activate the alarm if someone breaks a window and enters through the open hole. There are several methods of protecting glass in windows and doors. One is the metallic tape you may have seen on jewelry shop windows. It is sometimes circumvented by cutting around the tape, and many homeowners consider it unsightly. Another device is a vibration detector placed directly upon the glass. But, these devices may also be unsightly and may interfere with normal opening of the windows. Perhaps the best way to protect windows is by installing screens that are part of the alarm system. If such screens are removed or cut, the alarm is activated. But you may not want to use metal screens near salt water, since they deteriorate rapidly from salt corrosion, and you probably would not want to use screens on large picture windows.

In those cases you may want to install what are known in the trade as "interior traps." There are several kinds. One is alarm contacts on doors inside your home, say between the kitchen and a hallway. The idea is that, once having gained entry, a thief will walk around inside the house and set off the alarm by opening the contacted interior door. Another kind of trap is a pad under a carpet, which activates the alarm when stepped on. But, pets may also activate the alarm by stepping or jumping upon the pad. Another device, less sensitive to the weight of pets, is a stress detector that is placed under the floor joists of a house to detect movement of the floor.

Many people also use devices known as motion detectors. There are several types, and they, too, have limitations. An ultrasonic system generates high-frequency sound waves at a pitch higher than most people can hear. These waves fill an enclosed area. A receiver picks up the waves as they are reflected from walls and objects in a room. When a person walks through the room, that person's motion changes the frequency of the reflected sound waves, which the receiver detects. The receiver then signals an alarm control box, which activates the alarm.

One of the disadvantages of this system is that some women and many dogs can hear the high frequency pitch, which can be very irritating. Also, an ultrasonic alarm may not detect very slow movements, and it can be set off by loud noises, building vibrations, or air movements caused by an air conditioner or furnace, or any other movement in the room. Also, an ultrasonic system will only cover one room.

Microwave systems work on a principle similar to that of the ultrasonic system. A train of waves is produced that is partially reflected back to an antenna. If all objects within the room are stationary, the reflected waves return at a constant frequency, but if they strike a moving target, like a person walking, they return at different frequencies. The difference between the transmitted and received frequency is detected by the unit and activates an alarm. However, microwave systems can be difficult to control; they may penetrate the walls of the home and detect movement on the street or in the yard next door. They may also be activated by radio transmitters operating nearby.

Another motion-sensing device is a passive infrared detector. This device measures the amount of thermal energy in a room. If the pattern of thermal energy changes, as it will when someone enters the room, the alarm is activated. But, this device can be affected by thermal conditions outside the room. It is also limited to a line of sight, which means one can walk around its path or avoid it by moving behind obstructions in the room.

Perhaps the best known detector is a type you frequently see in small shops—the photoelectric cell. The system consists of a light projector and a receiver. If something breaks the light beam, an alarm is set off. Either white or infrared light may be used. But, these devices have disadvantages that limit their use in home alarm systems: They are easy to spot and avoid; the projector and receiver are easily moved out of alignment and then fail to function; and if the receiver gets dirty it will malfunction. A fourth type of system uses a relatively new device called an audio discriminator, which picks up the high-pitched sounds that usually accompany a break-in, like the sound of breaking glass.

Most motion detectors do not work well with animals in the house because the animals often activate the detectors. On the other hand, if you leave pets outside or with a neighbor when you are not at home, these systems can work reasonably well. But, if one of your goals is protection while you and your family are snuggled cozily in your beds, you will need to add more to your system than would be necessary in an unoccupied home.

What you have already installed will provide you with protection while you are in the house, but will also detect you and your family as you move around at night, just as an intruder would be detected. For that reason, you will need to zone your system, so that, for example, your doors and windows can continue to be monitored while allowing the occasional midnight snack or trip to the bathroom without alerting the neighbors and police to your nocturnal habits. Zoning allows you to turn your system on or off in a piecemeal fashion.

You must also decide whether you want to scare off the intruder with a bell or siren (illegal in some areas), whether you want the system to silently alert the police, or both. For home protection, I recommend a very loud, outside bell; you want to scare the intruder away, not catch him. The police are in the business of catching burglars. If you are concerned that the noise will offend your neighbors, use an inside alarm that will alert both you and an intruder to the fact the alarm has been activated, but won't disturb the neighbors. If you can afford it, your system should also be connected to a central alarm monitoring station or the police.

Even if you are not the owner of a jewelry store or bank, I also recommend a silent alarm, which alerts the police without sounding a bell. Such an alarm is usually activated by pushing a button. A good place to put one of these buttons is in the bathroom. If you are being held against your wishes, your captor will probably at least allow you the privilege of visiting the bathroom unaccompanied, and you can send the alarm from there.

So far we've talked about systems with components that are wired to an alarm control within the house. The same thing can be accomplished by using devices that are really miniature radio transmitters and report to an alarm control box by radio signal instead of wire. These devices may require frequent battery changes or some other form of backup power to operate the system in the event of an electrical failure, but they generally work well. One advantage of these systems is the avoidance of labor costs inherent in wiring a system, as well as eliminating the need to drill holes throughout your home in order to run wire. Also, you can use these devices on the patio, near the pool, or anywhere else within range of the radio receiver in your home.

If you use a bell or other outside alarm, get one that shuts off automatically after several minutes of sound. If the neighbors haven't heard it by then, they're not likely to. Those neighbors who have heard it are not going to be happy listening to it for long. Your alarm system can also be used to turn lights on, both outside and inside your home, when the alarm is activated.

If you install an alarm system, get used to having an occasional false alarm. Lots of things cause false alarms, but people cause most of them. There will be occasional mechanical problems, power failures, telephone line problems, and a host of other reasons why the alarm may go off when you don't want it to, including a "break-in" period (no pun intended) when your system is new and you are learning to live with it. Don't be discouraged by false alarms. An alarm system is composed of a series of electromechanical or electronic devices that will malfunction from time to time. Use your system. The more often you use it the less likely you are to cause false alarms. Train other members of your family to use the system properly, but don't train everybody on the block in how it works.

The police don't like false alarms. If you have too many false alarms, the police may even write you a citation, and if you have more after that, they may threaten to stop coming to your house when the alarm goes off. They politely call it "nonresponse status." For that reason, it will pay you to service your alarm system regularly. After all, you want it to work when you need it, and you want the police to respond to it. Test your alarm system monthly, and when you test it, test all of the activating devices. Your life may someday depend upon it.

Alarm systems can be set up to be turned off and on from either inside or outside the house by way of key, cardkey, or a digital pad similar to a touchtone telephone keyboard. You can have as many of these controls as you like, and you can have a mixture of types. Many are equipped with lights to tell you (and any would-be intruder) when the system is on and when it is not. Those with keypads are usually equipped with a button right in the keypad with which you can send an alarm if you are approached while you are entering your home.

Keep your system on whether you are inside or outside your home. The more you use it, the more comfortable you will be with it, and the more value it will be to you. A good alarm system is a major investment, and it earns no return if it's not used. A Scarsdale, New York, study showed that of all the homes burglarized over a six year period, only 2 percent had alarm systems and half of those weren't turned on when the burglary took place. (In the other half, the burglars entered through an unprotected opening.)

Before we leave the subject of alarms, make sure that you have smoke detectors in hallways between the bedrooms and the kitchen, living room, garage area, and anywhere else fire is likely to break out. Smoke detectors are a small investment compared to an intrusion alarm system, and you may realize a return on your investment more quickly.

Keep your alarm system as simple as possible, and be prepared for some frustrations while it is being installed and while you are getting used to it. If you are handy, you may be able to install your own system. They are now available in most hardware and radio parts stores, and security specialty shops are opening up around the country. Some systems require no wiring. Most do not connect to a central station. If you want a system that reports to a central station (and I encourage that if you can afford it), it's best to find a reputable alarm dealer in your community. Deal with the local office of a nationally known company or a local dealer who has been in business for a respectable period of time and has an impressive list of satisfied local clients. Don't just look at the list; check with some of the clients before you buy. In most communities today, you or I could go into the alarm business for the price of having a business card printed. Caveat emptor!

There is another type of system you might consider. It uses small microphones strategically placed in your home, with the sound level set to notify the central alarm station in the event of unusual noise. When an alarm is received at the central station, the people there turn on their receiver and listen to what is happening at your house. They are trained to detect the sound of fire, burglars, family members, and so on, and call the police, but I've always wondered what else they can hear. This kind of system is available from franchised firms around the country. One of its advantages is that false alarms are reduced.

If you don't want to buy an alarm system, you can lease one on a monthly basis. If you do buy a system and want it monitored, you will pay a monthly fee for the service and a charge for connecting it to the central station. You will also usually pay a separate monthly fee for a telephone line to the central station, and you may want to pay someone to maintain or repair the system if it malfunctions or breaks. If you want your system monitored by a company that also has guards who will respond to your alarms, you will pay extra for that service.

Alarm systems are getting better, but there is none that cannot be compromised. If you are going to use a central station, look for one that has a computer or microprocessor that provides line security. That means that if someone tampers with the telephone line or internal wiring connecting your system to the monitoring station, the tampering will usually be detected immediately. Unless you are in an unusual occupation or have unusually valuable items in your home, you will probably not be confronted with someone who will bother to tamper with your system in order to get into your home. Why should they when they can go to the home of a neighbor who doesn't have an alarm system? That's one of the basic values of *your* system. Many insurance companies have recognized the value of alarm systems and will offer discounts to homeowners who have them.

Because there is a growing market for security systems, more research is being done and technology is now rapidly being improved. There will be new, less expensive, more reliable alarm systems available in future years. But if you have need for an alarm system, don't wait. Select the best system for your needs and install it now.

5 Where There Is Smoke

As the old saying goes, there may be fire. This is a very short chapter. That's because this is not primarily a book about fire safety. But I include the topic for several reasons. First, I don't want you to do anything to protect your home that will inhibit your getting out of it quickly if there is a fire. Second, I feel compelled to remind you that more people are killed in home fires than are killed at the hands of intruders. We have one of the worst fire incidence rates of the industrialized countries. Fires kill about eight thousand people and destroy $6 billion in property in this country every year. Third, I want to say a few words about the crime of arson and what can be done about it.

When you consider locks for your doors, make sure they are of a type that enable you to open those doors from the inside without having to fumble for a key. Do not put bars on doors or windows that must be used by firemen to get in or by you to get out. Know how to call the fire department quickly. Hold family fire drills and include a designated place to meet outside your home to count heads after an evacuation.

If you suspect fire, alert others in the home and call the fire department. Follow the instructions in chapter 8 for dealing with a room fire. If you can't go out the door because it is hot, open or break a window and escape through it. If you're too high to jump, put a blanket over the window frame and sit there with one leg inside and one leg outside and wait for help.

Three-fourths of all multiple-death fires occur between eight at night and eight in the morning. A sleeping household has a 95 percent chance of escaping death or injury from fire once a smoke detector sounds, so use detectors liberally throughout your home. Install them between places where a fire may start, such as the garage, kitchen, rooms with a fireplace, and your bedrooms. In a two-story, basement home, put one in the basement, one on the first floor ceiling near the bottom of the stairs, and one on the second floor ceiling at the top of the stairs. If you can't mount detectors on the ceiling, mount them on the walls at least a foot from the ceiling and three feet from a corner. Keep detectors away from things that will activate them, such as drafts, fans, air ducts, the kitchen range, and the bathroom shower. Broiling fumes from the oven and steam from the shower may look like smoke to the detector, and it will sound the alarm.

Test your alarms monthly and vacuum the dust from them often. If they don't have a test button, blow a little smoke or incense their way to make sure they are working as they should. If your detectors are battery operated, change the batteries a minimum of every six months. In 40 percent of all fires, smoke detectors give the first warning, but, of those homes most vulnerable to fire, only a third are equipped with smoke detectors. The use of detectors will double your chances of surviving a fire.

If you replace glass in your windows or doors with nonbreakable material, remember that you won't be able to break out and firemen may have trouble getting in. Be certain you have an alternate escape route. You should always have two ways to get out of your home if there is a fire. If your home has more than one floor, rope or chain ladders kept in upstairs rooms may provide the only means of escape.

I grew up in a small community where members of the volunteer fire department were summoned by the wailing of the town siren. Consequently, our firemen did not respond to a fire as quickly as today's full-time fire fighters. For our own protection we kept garden hoses connected to the outside faucets at the front and back of our home for use before the firemen arrived. I still do that, and I recommend the practice to you. If you have a secondary water supply source, like a swimming pool, you can use that water to put out a fire, provided you have a pump. But never try to put out a fire by yourself before calling the fire department. Call them first. Then if you can safely extinguish a fire, do so.

Fire extinguishers put out 97 percent of the fires on which they are used, so keep the right kind of extinguishers handy near the furnace and in the garage and kitchen. But don't keep them too close to an appliance that might catch fire. You might not be able to reach the extinguisher when you need it. Water or dry chemical extinguishers will work on paper, wood, or cloth fires. Use a carbon dioxide or dry chemical extinguisher for gasoline, oil, or electrical fires. Fire extinguishers should be checked and serviced annually.

If you install your own alarm system or other security devices that require electricity, make sure your connections are properly made and that your fuses or circuit breakers can handle the additional load.

If your habits include falling asleep on the couch with a lighted cigarette or smoking in bed, you may have wasted your money on this book. Your life expectancy is less than average, and chances are good that a fire will strike you before a criminal will.

Most of us think of fires as something that occur accidentally or because of dumb behavior. But arson—deliberately setting a fire—is one of the fastest growing crimes in the nation and one that hits all of us in our pocketbooks. We pay directly, through loss of property, injury, or death, and we pay indirectly through increased insurance premiums and taxes that go to pay the cost of fighting these fires. The average arson loss is about $4,500, much higher than the $900 average burglary loss.

A Federal Bureau of Investigation profile of a typical arsonist points toward a white male under twenty-five, insecure and cowardly, unmarried, raised in a fatherless substandard home, and socially, sexually, and emotionally maladjusted. About 5 percent of arsonists are psychotic. Many burn to express their anger at someone or something. The rest have a variety of other motives. Arsonists burn for profit, revenge, spite, jealousy, sexual gratification, intimidation, vandalism, excitement, to conceal a crime, or perhaps because they suffer a mental disease known as pyromania. One man who rented fire fighting equipment was involved in setting four dozen wild-land fires to increase his business. Another burned dry grassland because he didn't like the color brown. The burning of businesses for insurance proceeds may be an organized crime endeavor or the work of an individual entrepreneur.

Arson can occur anyplace and can affect anyone. Even if you're not the target of the arsonist, if he sets fire to the house next door, yours may burn as a result.

What can we do about it? In addition to the things I've already mentioned to protect your home from intruders, clear your premises of old newspapers, leftover paint, old rags, and other trash that someone could use to start a fire. Secure flammable liquids, like gasoline for the lawn mower, in a locked, fire resistant cabinet, and never store it in the house. Keep storage areas secure. Dispose of all flammable waste materials as quickly as possible. Just as you would do if you saw suspicious people lurking in your neighborhood who you thought were going to burglarize a home, call the police if you see anything that makes you believe that someone is about to start a fire or already has.

In addition to increased fire insurance premiums and increased property taxes to support fire service, we suffer from arson through erosion of the tax base as property values fall in arson-prone neighborhoods. We suffer from loss of jobs at burned businesses and factories, loss of revenue to damaged stores and shops and a resulting loss of tax revenue, and inferior education facilities during reconstruction of burned schools.

Because nearly half of all arson fires are started by juveniles, a San Francisco program called Firehawks matches children who have set fires with firemen who serve as "big brothers." Funded with a grant from the U.S. Fire Administration the program so far has been successful. None of the youngsters in the program has set another fire since joining the Firehawks.

Fire is much less discriminating than a burglar or rapist who carefully selects his targets. Fire burns everything and everyone in its path. Protect yourself against fire.

6 Your Other Castle

Your home may be your castle, but you probably spend more waking hours in your office or workplace than you do in your home. The most frequent crime people face in their workplace is theft, either by outsiders or fellow employees. But there are some work locations that may subject employees to the risk of assault, and an employee who works in a retail establishment could be the victim of a robbery. Because of these risks, you should take time to review the security measures in place at your facility.

It is important to know what level of security is provided and not assume your workplace is safe. Employees often make that assumption because there are other employees around and because the employer has an implied interest in protecting the assets of the business. Since you are one of the primary assets of the business in which you work, no matter what your role in the organization, your employer's efforts should protect you. But the things the employer does to protect the plant, store, or office building may not protect you from being victimized by another employee.

It is not practical to try to provide specific instructions here for each work setting, but I can give you some advice that is applicable to most workplaces. A factory, industrial setting, or warehouse is often more secure than a downtown office building or retail establishment. Access to large industrial sites is often controlled by fences, guards, and electronic devices that limit visitors and thus reduce the chances of crimes committed by outsiders. Such facilities are usually well lighted and patrolled at night, so that employees going to and from their cars or other transportation are reasonably safe. Because there are usually other employees around, the opportunity for assault to occur, other than an occasional fight between employees, is limited. There is the opportunity for theft among employees, but in an industrial setting, employees are usually provided with lockers in which to place their belongings. Whatever kind of work space you occupy, personal belongings, particularly purses and wallets, should be locked up when you are not carrying them.

Retail establishments, of course, encourage visitors. While you want to get buyers in the door, nonemployees should not be allowed beyond the sales area, and employees should lock their personal belongings in suitable containers in noncustomer areas. If restrooms are provided for customers, they should be located in the sales area, not where they provide access to areas of the business where thieves can help themselves to your or your employer's belongings.

This book is not primarily concerned with the things employers should do to protect their businesses. For example, I won't be talking about bad checks, credit scams, and so forth, as they apply to businesses as victims. However, there is one crime that victimizes both you and the business—robbery. Many people confuse the terms robbery and burglary. Even newspaper reports can't seem to get it right these days. A robbery is a theft from a person of something of value, taken under threat or fear of bodily harm. The legal definition may vary somewhat from state to state, but common elements of the crime are threat or fear and the taking of something of value from a person against his or her will. Although many robbers are armed with knives, guns, or other weapons, some are not. For example, note-passing robbers commonly victimize financial institutions. Or a robber may simply demand money without displaying a weapon. Armed or not, the crime is still a robbery if fear or threat is used.

Since most robbers want cash, you can reduce your chances of being robbed by keeping as little cash around as necessary and by not displaying the cash you have. Take a tip from the convenience stores and gasoline service stations that are open late at night. Many are now equipped with safes that employees cannot open with slots into which money can be dropped as it is received. Those safes may periodically dispense small amounts of change by way of a time controlled cash dispenser. This system limits the amount of cash accessible to an employee at any one time. These businesses usually display signs that tell would-be robbers the system is being used.

Besides keeping your available cash at a minimum, make sure that the interior of any retail establishment is easily visible to passersby, particularly those areas where you make customer transactions and where the safe is located, and make sure the establishment is well lighted at night.

If, despite these precautions, a robber approaches you, be cooperative and do exactly what he or she tells you to do. While you have an obligation to protect an employer's assets, you don't have an obligation to be injured or killed while doing so. When you are confronted by a robber, your obligation switches from trying to protect your employer's money to protecting yourself while also observing every detail you possibly can that will help to apprehend the robber.

As soon as the robber has departed, sound the alarm, if the business has one, or telephone the police. Give them a description of the robber and the getaway vehicle, if you were able to observe it safely. Write the descriptions down as you repeat them; you will be asked to repeat that information over and over, particularly if the robber is caught and there is a trial, and you will be a more credible witness if you have recorded the information on something. Anything will do—the back of a napkin, a matchbook—but write it down.

If you work in an office building, you are less likely to be robbed, although it could happen if the firm for which you work deals in cash. In an office setting, the threat is usually theft. Most office buildings are open to the public and thus are attractive targets for sneak thieves, many of whom make a living going from one building to another.

Today's fire regulations make it nearly impossible to secure the floors of an office building. Fire codes require that people who find their way, legitimately or otherwise, to an elevator lobby must be provided two unrestricted ways to escape from the floor, not including the elevator. That means that you may find your work space accessible to any pedestrian who wants to visit your floor. Most businesses can't lock all their doors or customers couldn't enter, but most doors should be kept locked if fire codes allow it. Where doors are left unlocked, there should be a receptionist or other person whose job it is to screen visitors to the floor. If there isn't a receptionist, it becomes everybody's job to make sure visitors to the office are intercepted.

Many experienced office building thieves have agile tongues. If you find them in your office area, they may tell you they are looking for the restroom or the personnel office, or they may give you the name of a nonexistent person they are there to see.

A frequent modus operandi of the office building thief is to steal wallets from men's jackets, which a thief can frequently find hanging behind an office door, or from women's purses, which are often found on the floor by a desk or typewriter. A thief will often take only cash, checks, and credit cards, and leave the wallet or purse behind. It will take longer to discover the theft if the wallet or purse are left behind, and the thief does not risk getting caught with that incriminating evidence. Using your credit cards, the thief may cash several of your personal checks within an hour of relieving you of them. If you have not gotten the point of this scenario, I'll be more direct. Your wallet should either be in your pocket or locked in a safe place, and your purse should be on your arm or locked up.

If you must leave valuables in your office, secure them somehow when you leave. Many people have expensive desk sets, clocks, personal radios, calculators, and other treasures in their offices. It's best to leave those things at home. If you must have them in the office, place them in a desk or credenza that locks, lock it when you leave, and take the key with you. Don't leave keys under the desk blotter, in the unlocked desk drawer, or under the letter "k" in the rolodex. Most thieves are not dumb, or they would be caught much sooner. They know where you hide your keys, as does everyone else in your office. So don't hide them. Take them with you. Make sure your office door has a lock on it, and lock it when you leave.

An acquaintance, whom we shall call Nancy, visited a client in the client's downtown office. When Nancy arrived at the reception area, there was no receptionist present. Being a patient sort, Nancy helped herself to a seat and waited. She was soon joined by a well-dressed couple who asked the whereabouts of the receptionist. Nancy replied that the receptionist must have stepped out, since she hadn't seen her. The couple then casually entered the office area, leaving Nancy behind. As they meandered through the office, the couple was approached by an employee who asked if she could help them. They explained that they were looking for a list of the company's branches, so the helpful employee went to her office and returned with her copy of the list. The man then asked if he might use the restroom. The restrooms were kept locked for security reasons, but the helpful employee went in search of a key, found one, gave it to the man, and showed him to the men's room. That was the last our characters saw of the couple. And the cash and credit cards that were taken from a manager's office during this episode weren't seen again by their owner. The checks showed up at the bank. Several had been cashed by the thieves. And one of the credit cards was recovered from another thief who was attempting to use it in a local store.

If you see people you don't know in the vicinity of your office, ask if you can help them. If they're honest, they'll appreciate your help, and if they're not, you don't care about their feelings. Never let strangers out of your sight, or make sure someone else watches them until they have left your office. If they refuse your assistance but aren't leaving fast enough to suit you, call your building security office, the police, or whatever other help is handy. Don't attempt to eject an intruder by yourself unless you are exceptionally skilled in the martial arts.

Both the skyrocketing cost of office equipment and the long lead times for delivery make certain pieces of equipment extremely attractive to a thief. Thousands of electric typewriters are stolen from office and commercial buildings in this country each year. One firm for which I worked was plagued with such thefts, both from office buildings and from branch offices. The modus operandi was slightly different for each. Branches were the regular victims of amateurish but effective burglaries. Glass doors or windows would be smashed with whatever was handy, three typewriters would be grabbed, and the thieves would be gone. We puzzled over the thefts of three typewriters at a time, but then realized that three fit comfortably in the trunk of an automobile.

We pondered the office building thefts. No amount of increased security seemed to help us or our neighboring office buildings, which were experiencing similar thefts. Although we weren't sure how the typewriters were being taken, we decided to bolt all of our typewriters to the desks they sat upon. The thefts stopped overnight. We had increased the time necessary to steal the machines, and the thieves perceived an increased risk of being caught.

Although our thefts stopped, they did not stop in neighboring buildings where typewriters had not been bolted to the desks. One evening an alert guard spotted a former employee wheeling mail bags out of a building on a hand truck. The bags contained not mail but typewriters, and the thieves went to jail. They had previously worked in several of the buildings and had retained their identification cards upon terminating their employment. Using the old ID cards and posing as messengers, they plied their illegal trade.

From the point of view of your personal protection, the most vulnerable areas of an office building are the fire stairs, the restrooms, the elevators, and the parking garage. Fire stairs should be used only in an emergency. If there is a fire and you use the stairs, you will probably have a lot of company. At other times, you will usually find the stairs deserted, as they should be. Fire stairs should always be locked from the stairwell side, so that intruders cannot use the stairs to travel from floor to floor, and stairwells should be equipped with alarms that signal building security personnel when someone enters them. That way, the stairwells can't be used as a place to drag a robbery or rape victim. Instead of campaigning to have the stairwell doors left open so you and your friends on the floors above and below can run up and down to see one another, you should campaign to keep them locked.

Restrooms should be for the use of those who work in the building. Depending upon the occupancy of your building, the area in which it is located, and the kind of business conducted in it, it may be wise to keep restroom doors locked. When you are in a public restroom, don't leave your purse or briefcase on the hook on the cubicle door or on the floor near the door. It is relatively easy for a thief to reach over or under the partition to take your purse or briefcase. You are going to be somewhat hampered by the fact that you are sitting down with portions of your clothing wrapped around your legs. Hot pursuit of the thief will be difficult and probably embarrassing.

In the days when bombings of buildings were more frequent in this country, restrooms were a favorite hiding place for the bombs. There were two reasons for that. Restrooms are easily accessible when other areas of a building are not, and a bomb exploding in a restroom gave the bomber a bonus. Not only was there damage done by the initial blast, but an explosion in a restroom broke water pipes, and the water created damage on other floors of the building.

Especially during periods of low occupancy of a building, restrooms are a good place to surprise someone, or to drag them into for less than honorable purposes. If the restrooms in your building are left unlocked and your building is a public one, there may be times when you will want to use the buddy system.

Particularly when working after hours, be careful about the company you keep in an elevator. If you don't like the looks of someone in the elevator, don't get in. If you're already in when they get in, get out. Elevators are usually equipped with alarm buttons for use in an emergency. Don't hesitate to push one if you're in trouble.

If you park in the building garage, try to leave when others are leaving. If you must work late, try to accompany someone else when going to the garage. Have your keys ready, and look into your car before you get into it to make sure you don't have uninvited guests.

Bomb threats are a popular sport aimed predominantly at large businesses. The motives of a person who calls to say that a bomb has been placed in your building may be complex. He or she may be angry at the company or an employee for any number of reasons, may be attempting to extort money, or may be an employee who simply wants the afternoon off. At any rate, there is little correlation between bomb threats and the actual detonation of an explosive device.

Of course, the mere mention of a bomb strikes fear in the hearts of most people. But the fact is that few true bombers will call first to tell you their plans, and bombings in this country have decreased substantially in recent years. Figures from the Alcohol, Tobacco, and Firearms Bureau of the federal government show there was a 38 percent decrease in bombings in the United States between 1975 and 1980.

The Civilian Intelligence Agency reported 1,774 terrorist bombings in the world between 1968 and 1979. Only 229 of those were in North America, and only 80 of those were directed against United States companies or citizens. Finally, residences, not businesses, are the most frequent targets of bombings, and vehicles are bombed as often as businesses. If you're interested, in order of frequency, other targets are schools, mailboxes, open areas, utilities, government offices, banks, police stations and cars, military bases, and aircraft.

But what should you do about a bomb threat at your office? Most large firms have a "bomb-threat checklist" upon which the receiver of such a call records exactly what was said by the caller, characteristics of his or her voice, and background noises. Once that is done, the building manager, building security, and the police should be notified.

Bomb threats are so frequent that most police agencies simply do not have the manpower to conduct a search every time there is a threat. In one week in New York City, there were hundreds of bomb threats. Don't expect the police to roll to your bomb threat with the bomb squad and dogs trained to sniff out explosives. While television cops do that, in real life you'll get the police officer on the beat who will call the bomb squad only when an object resembling a bomb has been found. If something unusual is found, obviously you should not touch it and should keep others away from it until the police have an opportunity to examine it.

The police probably won't offer advice about whether or not your building should be evacuated. That decision will be left to someone within the company. Before making such an evacuation decision, whoever makes it should consider the probability of a bomb having been placed in their business. There are few bombings in this country, most are not directed at businesses, and most bombers don't call in advance. Many businesses have access control procedures or are located in premises that would make it difficult to place an explosive device without being observed. So, most bomb threats are spurious. But, you say, isn't it safer to evacuate, just in case? The policy of evacuating every time there is a threat encourages more threats, and every evacuation upsets the employees and customers of a business. Eventually, if calls persist on a frequent basis, a company must make a decision to stop evacuating or close its doors. So a careful evaluation of the threat and the probability of there being an explosive device on the premises makes more sense than an emotional reaction to evacuate the building.

If an evacuation is going to be made, it should be done in accordance with a well-thought-out and previously announced plan of action. You should know how to get out of your building quickly and safely. You should know where emergency exits are located, how an evacuation signal will be given in your building, where you are to assemble once outside (to count noses), who will assure that handicapped people will be helped out of the building, and how the signal to re-enter will be given. An orderly evacuation process will reduce confusion, avoid panic, and perhaps save lives.

Large firms usually have group or floor wardens whose responsibility it is to see that everyone leaves the building safely. Many provide written materials and training programs that include diagrams of the building or floors, with the exits clearly marked. Depending on the type of emergency, you may not want to use the elevators. For example, fire and earthquake are emergencies that dictate the use of stairs, not elevators, for evacuation.

As a general rule and despite the opportunity for petty theft and other annoyances, you're probably safer at work than you are elsewhere, and you can always decrease your odds of being victimized by being alert to the more common hazards of the workplace.

7 Wheeling and Dealing

Your vehicle may be the most substantial piece of property you own, and you may feel very secure while you are in it. Most vehicles are made of steel, they keep the rain out, and move you from place to place. You can sleep in a vehicle, socialize in one, even live in one. Your car can serve as a protective barrier provided you follow some basic rules.

Always keep your vehicle in good running order, with the fuel tank never less than half full, so that you minimize your risk of being stranded somewhere. If you don't know how or don't care to service your car yourself, have it serviced regularly by a dealer you trust.

You really must know how to drive your vehicle. While that sounds very basic, perhaps even silly, what I mean is to know, through experience, how your car will respond in an emergency. Will it accelerate rapidly when you need to get away from someone or something? How quickly will it turn when you need to avoid a situation? Do you know its turning radius in the event that you need to make a midblock U-turn to avoid trouble? Are you a defensive driver who can steer around trouble, or do you simply apply your brakes in an emergency? Are you constantly watching the front, rear, and sides of your car as you drive, so you can see trouble if it approaches? When you are stopped in traffic, do you leave enough space between you and the car ahead so that you can pull around that car if you need to get away quickly? These are defensive driving techniques that are not usually taught in a driving class, but you need to know how to do these things instinctively and intuitively.

When you drive, always keep your doors locked, front and rear, to prevent an uninvited guest from popping into your car with you, and keep your windows rolled up. Convertibles are in vogue again. If you should be so fortunate as to own one, be careful when and where you drive with the top down. If someone tries to flag you down in what looks like an emergency situation, roll your window down an inch and talk through the opening. Offer to go for help, but avoid getting out of your car unless the police are present. And never stop for hitchhikers, no matter how innocuous he or she may appear.

Park as close to your destination as possible. At night, park in well-lighted areas. Don't get out of your car if things don't feel right or if there are suspicious people around. Lock all your doors when you park, and be sure to lock valuables in your trunk. Even the bears in Yosemite National Park have learned that the blanket on the back seat may be covering something of interest. The glove box is no place to leave valuables either. That's the first place a thief will look. When you return to your car, look inside before you enter to be sure unexpected company is not waiting for you. Have your door key ready so you can enter your car quickly.

If you leave your key with a parking lot or garage attendant, leave only the ignition key and not your trunk, house, or office keys. Keys can be copied and used by someone who desires to visit your home while you are away, so don't put your address on your keys. If you lose them, it's better that no one knows whose locks they fit. If you keep a garage door opener in your car, lock it in the trunk or take it with you. It could be used to enter your garage in your absence.

Avoid the use of vanity license plates, which easily identify you. It's best to be as anonymous as possible while on the road, particularly if you are earning a high income or are in an occupation that may target you for kidnap or extortion. Optional but valuable equipment to prevent tampering with your car includes a locking gas cap and a hood lock.

It's wise to carry certain accessories in your car: a flashlight, highway flares, a tow chain (for towing your car, not someone else's), a first aid kit, and maps of the areas in which you normally drive or plan to visit. A jack, spare tire, and lug wrench are not accessories but necessities. A handy optional device is a small, folding shovel, useful for digging out if you get stuck.

Some sort of two-way communications is also a good idea. Citizen's band radios that plug into the cigarette lighter and with an antenna that sticks magnetically to the roof are available. The advantage of this type of CB is that it does not have to be permanently installed in your car, but is available for use in an emergency. CB radios have a very limited range, and fewer people monitor them than in the past, so they may not be useful in remote areas. If your budget is up to it, you might consider a car telephone, which is really a radio telephone. It will provide wide coverage, even in most isolated areas. CBs, telephones, tape decks, and similar accessories are very attractive to the thief, so it is a good idea to install them where they can be used but not seen from outside the vehicle.

You may be able to conceal a telephone in the glove box or a CB in the center console. Concealable antennas that are mounted in the rear of the vehicle are now marketed, so you don't have to give away the presence of your electronic equipment by displaying your antennas. Many insurance companies will not cover the loss of CBs or tape decks without the purchase of additional coverage. That coverage may be worth the cost.

At least one U.S. auto manufacturer is offering a driver security package option on its new models. It features an emergency road kit with first aid supplies, an audible alarm, battery booster cables, road flares, a fire extinguisher, screwdrivers, a gas siphon and a flashlight with a safety cone. Other features include a CB radio, a long-life battery, puncture-sealant tires, power windows, power door locks, and a rear window defroster.

There are those who advocate the installation of an alarm system in your vehicle, and some cars come from the factory with alarm systems already installed. There is no question that cars are frequently burglarized or stolen. Vehicle theft costs American consumers about $6 billion a year and not just in the value of stolen vehicles and parts. City dwellers pay $200 a year more for automobile theft insurance than country folks because of the added risk of theft. One of every forty-three registered vehicles in this country will be stolen or broken into this year and every motorist stands a 30 percent chance of having a car stolen during his or her lifetime. More than a million vehicles are stolen each year, an average of one every thirty-two seconds.

Forty percent of today's stolen vehicles are stripped for parts, since the individual parts often sell for more than a complete car and are much more difficult to trace.

The best defense against the theft of your vehicle is to use common sense when parking it and to make sure you lock it when you leave it. Since 60 percent of car thefts happen directly outside the owner's home at night, lock it when at home as well. You can install auxiliary locks if you are uncomfortable with the standard ignition and door locks. One type of auxiliary lock anchors the steering wheel to the brake pedal. A hidden "kill switch" cuts off the car's fuel supply or electrical system. Some ignition control devices require the entry of a code to start the car.

Dozens of alarm systems are available, enough different types to boggle your mind if not that of a criminal. There are systems that activate when the car is jiggled, those that respond to the frequency of breaking glass, those that have hood and trunk contacts, and those that monitor the current in the car's wiring system and detect when the courtesy, hood, or trunk lights come on. If you use an alarm system, install a hood locking device so the thief can't open the hood and disconnect the power source, or install an alarm system that has an independent power source.

Some alarm systems activate horns or sirens (illegal in some communities) that can be heard miles away, some systems send a signal to a pager in the owner's pocket, some systems flash the car's headlights, and one system will even yell ''burglar, burglar.'' My experience is that few people pay attention to a vehicle alarm, except those who are annoyed by them. I don't recommend them for most vehicles. Like other alarm systems, these systems will go off when you least want them to, for no obvious reason, and the alarm will frequently continue to sound until the battery dies or a disgruntled passerby disconnects the system. At a local church recently, a driver parked his automobile in the parking lot and left on a church field trip for the day. For unknown reasons the alarm went off, disrupting a wedding taking place inside the church. The police were called and had three options: Allow the alarm to continue to sound, disrupting the wedding; open the car and disconnect the alarm system, leaving themselves potentially liable for leaving the car without protection; or impound the car as a public nuisance. Since I didn't stick around while they decided, your guess at the outcome is as good as mine.

On the other hand, if you have a recreational vehicle in which you sleep while you are on the road, I recommend an alarm system. A vehicle system or a variation of those discussed in chapter 4 can be adapted to a motor home, trailer, or camper.

Whenever you travel in unfamiliar areas, plan your routes in advance. There are places where you may not be welcome, and you should know that and avoid those places. Check with your automobile club or travel service, or plan the trip yourself, but plan carefully. Stick to freeways and major thoroughfares whenever possible.

If you do get into trouble while in your car, flash your lights or use the horn to sound the alarm. If you are being attacked, the noise alone may drive off the culprit. If you are being followed, drive to the nearest open business, firehouse, or police station and notify the police. When I was younger, parents warned children to stay away from lovers' lanes. I don't know if they're still called that or if people still frequent them, more convenient places being available, but the advice is even better today than it was years ago.

Replace the door-lock posts in your car with the type that cannot be lifted with a coat hanger. That will make it much more difficult for someone to break into your car, though it will also make it difficult for you if you lock your keys in your car. Always take your keys out of the ignition when you turn off the engine, and put those keys in your pocket or purse. Should you hide an extra key somewhere on the car? You decide, but remember that most of the places you will think to hide the key will also be thought of by a thief who is looking for it.

If you have car trouble, get out, raise the hood, get back in, and lock the doors, unless your car is disabled in a hazardous position where it is likely to be struck by another car. Then you have an obligation to try to warn other drivers. That's the reason for the highway flares you should always carry. If you've never used a flare, practice in advance. They don't light like anything else you are used to lighting. They will burn through your clothing and skin if you don't handle them properly, and they can easily start brush and forest fires.

While you have an obligation to try to warn others if your car is stuck in a hazardous position, you don't have an obligation to get killed doing it. Stay out of the traffic lanes and do only what is necessary to signal others. If you want to learn to be a traffic director, join your local police reserve, but don't begin on the freeway. If you must abandon your car, keep all passengers together.

If you drive in the colder areas of our country, don't panic if your car fails to start in some remote area or gets stuck in the snow. Your chances of survival are better if you stay in your vehicle, even if it is buried in snow. Exercise to keep your circulation and body warmth up, and if you leave your car to do that, stay within sight of it.

Try to stay awake if you are alone. If you have company sleep in shifts. Run the motor every half hour or so to generate heat, but conserve fuel by not running it continuously. Leave a window open slightly for ventilation, and keep the exhaust pipe clear of snow so that carbon monoxide fumes won't seep into your car. Turn the inside dome light on occasionally at night to signal your presence, but be careful to not use all the "juice" in your battery. Ration your food and drink (the snow will provide lots of water) and avoid alcoholic beverages. They will reduce your body's resistance to cold. If you travel often in the wintertime, carry blankets, an ice scraper, a brush or whisk broom, and tire chains with you.

Various devices exist that you can use to signal for help when your car is disabled. One emergency sign kit has a holder that sticks to the top of the car with magnets and includes different signs that indicate the need for gas, water, police, or just plain help. There are inflatable devices that look like traffic cones. Some auto clubs furnish free signs that you can display in time of need, and a rag or piece of clothing tied to the radio antenna is a universally understood signal of distress.

If you have a flat tire while on the freeway or in a less than desirable neighborhood, drive slowly, on the edge of the roadway, to a safe spot to fix it. Chances are the tire may already be ruined, and a tire is cheaper and easier to replace than you or the remainder of your vehicle.

Most of us will never be targets of a kidnapper or terrorist and so do not need to take special precautions against them. But there are various schools around the country that teach you what I call "counter-offensive" driving—driving to get away from the would-be terrorist. The courses are expensive, but they are valuable for a select audience. They teach you how your vehicle will respond at high speeds and in tight turns, how to drive offensively, and how to use your vehicle as a defensive weapon if you are attacked.

Some people like to play traffic cop when they drive. Those who are righteously inclined will drive fifty-five miles an hour in the fast lane, no matter how many cars stack up behind, to insure that everyone else stays within the speed limit. Some like to chase errant drivers down the road, beating on the horn. Others like to bully another driver off the road in order to give him or her a piece of their mind.

Leave the driving instruction to someone else. Worry about your own driving, and if someone else makes a mistake, let yet another person correct it. If you make a mistake and are rewarded by obscene salutes, yells, or horn honking, put on your best sheepish grin and go on about your business. The streets and highways are no place to argue, and playing bumper cars with full size, fully loaded vehicles is both dangerous and expensive.

There is a possibility, but low probability, that someone will deliberately involve you in a traffic accident in order to rob you or commit some other crime. If that happens, the "accident" will likely be a minor one designed only to get you to stop your car and get out of it. If you are involved in such an incident and are suspicious that it was deliberately caused, stay in your car and talk through the window. While you are required to exchange information with the other driver, you can do that by rolling down the window an inch or so and exchanging slips of paper. Don't get out of your car unless the police are there. If things really look bad, you can drive to the nearest police station and make a full report of the incident. You may run afoul of laws regarding hit and run driving, but that is unlikely if no one has been injured and you go directly to the police.

All in all, driving should be a rather safe activity, but, like everything else, it has its hazards. Unfortunately, the danger doesn't always come from those who don't know how to drive.

8 Traveler's Aid

To many, traveling is a recreational or vacation activity; to others, it's a necessity of business or life. Most of us are susceptible to certain risks while traveling. Before you travel, do as much advance planning of your trip as is practical. Know where you want to go—both the city and the destination within it—and how you are going to get there. Even though you may fly to your destination, you will be more comfortable in a strange city if you look at road or street maps before you go since you will need ground transportation after you arrive at the airport. It's useful to know how to get to your ultimate destination if only to prevent a greedy cab driver from giving you a grand tour without your knowledge or consent.

Don't leave your baggage unattended at an airport. Use solid luggage and lock it. Most people think their baggage is safe once it has been given to the skycap or checked at the airline ticket counter, but valuable items have been known to disappear from suitcases between check-in and arrival at the destination city. Carry expensive jewelry in your carry-on luggage, if you are going to wear jewelry while on your trip. Otherwise, leave it and any other expensive item you won't be using at home or in your safe deposit box.

The airlines require that your name be on the outside of each piece of luggage you check, but that could allow a dishonest airport employee to record the names and addresses of those who will be out of town for a while, and that information could be used to decide which homes to burglarize. To guard against that possibility, use your business address on your luggage tags. When you get to your destination, claim your luggage right away before others have an opportunity to claim it for you.

If you park your car at the airport, thieves may be able to determine that your home is temporarily empty. They may obtain your address by checking your license number through the state motor vehicle department. Your car is also subject to burglary or theft when you leave it for long periods in a public lot. You may be better off to take a cab or other public transportation to the airport, or have a friend or relative drive you.

If you are driving to your destination or will rent a car upon arrival, plan your routes carefully. There are areas in almost every city where you are not welcome. There are other areas where you are as welcome as any chicken who stumbles into the den of the fox, and you do not want to end up in one of those areas by mistake. When driving, don't advertise that you are a tourist; put maps, travel guides, cameras, and binoculars out of sight. When unloading your luggage at a hotel or motel, don't leave the keys in the trunk lock or your luggage unattended.

Before you leave home, make sure your trip plans, itinerary, and emergency telephone numbers have been left with a trusted friend or neighbor. Someone should always know where you expect to be, when you expect to return, and how to reach you in an emergency. Don't advertise your vacation in the society pages of your local newspaper, and don't set your baggage in front of your house for passersby to see that you are leaving.

Pick a well-known, reputable hotel or motel in which to stay. There's no guarantee it will be crime free, but there will be a better chance that your hosts will not be part of the criminal population. If you are not familiar with hotels and motels, make reservations through an established travel agency in which you have some confidence, or through one of the many travel services offered by automobile clubs or gasoline companies. When you register, use your last name and your first two initials to conceal your gender. If you plan to arrive late, make sure you guarantee your room for late arrival, so it's still there when you get there.

Consider the plight of the unfortunate visitors to one of our country's world fairs who sent their reservation deposits to a third-rate hotel run and occupied by thieves. When the visitors arrived and found their accommodations poor, they had no other options, since they had paid for their rooms in advance and no other lodging was available. They reluctantly accepted their rooms, vowed to make the best of it, and departed to see the fair. When they returned all of their belongings worth stealing were gone. Despite the fact that the hotel operators were well known to the police, there was no proof that the operators had participated in the thefts. Try suing a crook sometime. Most have few assets.

If you are attending a seminar in a strange city, it is safer and more convenient to stay in the seminar hotel. While it may be slightly more expensive, you will probably save the difference by avoiding transportation costs between hotels. By staying in the seminar hotel you reduce the number of times you must venture into unknown territory. If you do intend to explore on your own, ask at your hotel or motel for the safest route to your destination, and inquire about any areas you should avoid. Although most cab drivers are honest, you should use only licensed cabs operated by well-known companies. While you ride, don't reveal personal information, and act like you know where you are going and what you are doing.

If you walk, stick to well-traveled main streets and public areas, and, if walking at night, those that are well lighted. If you get lost, find an open business that looks reputable and ask directions. Try not to look lost. Avoid standing on a street corner perusing your map. Looking bewildered gives you away as a tourist and increases your vulnerability.

You should, of course, limit the amount of cash you carry and use traveler's checks whenever possible. Keep a record of your traveler's checks and credit cards with you, but don't keep it in the same place you carry your cards and checks. Don't carry more credit cards than you will need. Have with you the telephone numbers to call in the event your cards or checks are lost or stolen. If you must carry a large sum of money, don't carry it all in one place. Put some in your wallet or purse, some in one pocket, the rest in another. If you are robbed or your pocket is picked, you may not lose everything.

Many cities have street merchants and entertainers who add to the local color. Enjoy them, but be aware that some may attract a crowd for the sole purpose of assisting the pickpockets with whom they work. If they are not in cahoots with the crooks, they may still unwittingly aid them by attracting a crowd of people whose attention is diverted from their belongings. People who bump or jostle you may be pickpockets. Keep your wallet in an inside coat pocket and carry your purse securely, a subject I'll cover in chapter 11.

Although we don't hear much about pickpockets in the news, from 1976 to 1980 the rate of these crimes increased nearly 50 percent in the United States. Today pickpockets are well trained. Those trained in a South American school learn to empty the pockets of a mannequin without disturbing a bell sewn in each pocket. Nary a tinkle is to be heard, or the student will not graduate.

Pickpockets often work in specialized teams. Some work streets, others buses or subways. A "stall" may drop money on the street to distract the "mark." When the mark (you or I) reaches for the money, the "pick" steals the mark's wallet and gives it to the "handoff," who takes off. The handoff will take the money out and drop the wallet in the nearest trash can. The stall or the pick, if caught, will have no incriminating evidence upon them.

Pickpockets sometimes use more basic techniques in their trade. In Los Angeles, pairs and threesomes of female pickpockets distracted their elderly male victims by baring their breasts; they also plied their pigeons with intimate caresses. Airport pickpockets often watch ticket counters or newspaper stands to determine the location of the mark's wallet. They work the ticket lines and the area just outside the security screening area. It's the departing passengers, who still have plenty of money in their pockets, that they're after.

All this has encouraged a market for antipickpocket devices. One device is a case about the size of a cigarette package with a cord, battery, and buzzer. The case goes on the belt, the cord is attached to the billfold, and if the billfold is removed, the buzzer buzzes. Another device fits inside a wallet, stretches it out, and makes it harder to remove from the pocket. An inexpensive deterrent involves attaching Velcro to the wallet and the pocket to make it more difficult for a thief to remove the wallet from the pocket. Most people don't need these devices.

Beware of strangers who go out of their way to befriend or assist you, especially if you have not asked for that attention. The fellow who offers to arrange companionship for the afternoon or evening may provide you with more excitement than you had planned. In the case of those who divest you of your money up front for such entertainment, you may find less excitement than you had hoped for. A guy who offers to sell you something cheap probably knows what his merchandise is worth.

When you meet strangers, don't reveal the length of your trip, where you are staying, or the fact that you are traveling alone. If you meet someone you'd like to know better, get his or her telephone number. Arrange a first meeting in a public place, not in a hotel room or in their home.

Remember that commercial lodging is designed primarily for sleeping and not for protection. When you register, find out how your hotel notifies guests if there is a fire or emergency. According to the United States Fire Administration, there are some twelve thousand fires in the nation's fifty-two thousand hotels and motels annually. Careless smokers are responsible for about a third of them and arsonists for about 15 percent.

There were 370 serious fires (those resulting in death, injury, or damage exceeding $100,000), 849 deaths, 2,645 injuries, and more than $177 million in damages in the United States lodging industry between 1971 and 1981. While that sounds bad, the average is only 37 serious fires a year in more than 50,000 motels and hotels which have 2.5 million rooms. And while there were an average 85 deaths per year, that's slight compared to the 50,000 people who die each year in traffic accidents.

Nevertheless, because of recent hotel fires in buildings considered to be modern, most major hotels are sensitive to the problem and have taken measures to educate their guests in fire safety. You may find a fire safety brochure in your room, or you may even be handed one as you check in.

While being shown to your room, scout two escape routes from it. Determine the most direct route to and from the room, elevators, fire escapes, and nearest telephones. Mentally plan your evacuation route. Some suggest you count the doors between your room and the emergency exits. How helpful that will be depends upon your memory for such things. Locate the fire alarm box and fire extinguisher closest to your room. Read the instructions so you know how to use them if the need arises.

When you enter your room, put your key where you can find it in an emergency. Then take a look at your room ventilation. Many older hotel rooms take in "fresh" air from the hallways. When there is a fire, the smoke from the hallways is sucked into the guest rooms. If there is a fire outside your room, and you can't get out, you will want to cover that vent. Locate the fan or air conditioning controls in your room so that you can turn them off if there is a fire. Check the windows and doors to see how they operate. See what's outside your room window. Is escape possible that way?

If there is a fire in your room, call the hotel operator or the fire department and tell them the location of the fire, or activate the fire alarm. If it is a small fire and you can put it out safely, do so, but only after reporting the fire. If not, give the alarm and escape. If you're in bed and there's evidence of a fire, roll out of bed and crawl to the door. Smoke and deadly gases rise, so don't stand up.

If you think there is fire outside your room, feel the inside of your room door. If the door or knob is hot, do not open the door. If it is cool, open the door slowly, but be prepared to slam it shut if you find fire on the other side. Walk to the fire stairs, taking your key and locking the door behind you. If there is smoke in the hallway, and you must reach the stairs, crawl. Stay close to the wall so you do not become disoriented.

Do not use the elevator. In modern buildings, elevator cars return to ground level when a fire alarm is activated. From that point on, they are controlled by arriving firemen. In older systems, the elevators may be stuck on the fire floors by smoke which, passing by the electric eyes in the doors, holds them open. If you think you sometimes wait a long time for an elevator, imagine waiting for one while the building is burning. When you get to the fire stairs, feel that door before opening it. If it is hot, you may have no choice but to return to your room.

If the door is cool, enter the stairs, close the door behind you, and walk down and out of the building. If you encounter smoke while walking down the stairs, you may need to walk up the stairs to the roof. Wait there to be rescued.

Depending upon the type and location of a fire and the construction of the hotel, you may be safer if you stay in your room. If there is smoke or fire in the hallway, close the door and shut off the fans or air conditioning. Some fire departments recommend you place the "do not disturb" sign on the outer door knob to signal arriving firemen that the room is occupied. Soak towels, sheets, and blankets with water and stuff them in cracks around the door and air conditioning vents. Remove the draperies from your room window. If fire from an adjoining room should break your window, the draperies may ignite. Fill the bathtub, sink, and any other container with water with which to keep the towels and linens wet and to douse any spot fire that may break out in your room. Don't break a window if there is smoke outside—you may just suck that smoke into your room. By telephone, let someone know where you are. Remain as calm as possible and wait for help. If you are on the first floor or second floor, you may be able to safely drop to the ground from your window. A drop from greater heights usually results in injuries and sometimes death.

If there is no smoke or flame outside your window, you can hang a sheet outside as a signal to firemen. You can also make a tent over your head with a blanket and lean out of the window to get fresh air, but only if there is no smoke, heat, or flame outside.

A chute is now being marketed for escape from high rise buildings. There are interior and exterior models, and even a mobile rescue chute. The chute is constructed of a double sleeve of nylon and will withstand heat up to 800 degrees centigrade. Evacuees climb into the chute and control the speed of their descent by flexing their arms and legs. The manufacturer says the chute can accommodate up to thirty people per minute. While new in the United States, it has been installed in more that fifteen hundred buildings throughout Europe. I've seen it demonstrated, and it works.

Relatively inexpensive, self-contained smoke masks are available. Although usually installed in containers inside buildings, there's a portable model you can carry with you when you travel.

If you are nervous about fire in a hotel, insist on being assigned a room on the first seven floors. Most fire departments have aerial ladders that will reach that high. If your room does not have a smoke detector, you can buy your own portable version to take along with you. It will cost less than a third of a night's room rent and could save your life.

Anytime you are in your hotel room, use all of the locking devices available. I've been awakened in the middle of the night by someone checking into my already occupied room, and at one major Los Angeles hotel I was given keys to three different occupied rooms before I was given a key to one that was empty. Don't leave the door open while you sneak down the hall for ice. When you reach your room, unpack and place your belongings in the closet or dresser. Arrange them so you will know if something is missing. Lock your empty suitcases so that, if someone does get into your room, they can't use your luggage to carry the rest of your belongings out of the hotel. Make sure you have identification inside your luggage as well as outside.

Like the portable fire alarm, there is a portable burglar alarm you can take with you when you travel. It takes little space, hangs on the inside door knob, and is powered by a battery. When someone touches the knob on the other side of your door, your alarm will sound. It will not discriminate, however. Whether the room service waiter or a burglar touches the door makes no difference to the alarm; it will make the same sound.

Never leave money, checks, credit cards, jewelry, car keys, or airline tickets in your room. Airline tickets are negotiable. Thieves sell them at discounted rates for immediate cash. You, however, will have to wait months to receive credit for your lost or stolen ticket and may not get a refund at all if it has been used by someone else. Place your jewelry and excess cash in the hotel safe or in safe deposit boxes provided by the hotel for that purpose.

Take a good look at the key your hotel hands you. The smartest hotel operators don't put the hotel name on their keys. That way, if a key is lost or stolen, the finder or thief can't use it without researching which hotel it came from. Reputable hotel owners are spending hundreds of thousands of dollars on ways to prevent hotel room crimes. Some have installed programmable card access systems that can be changed every time a guest checks out of a room. Others have gone to easily changeable lock cylinders to lick the lost or stolen key problem.

Always leave your key in the same place when you are in your room and never leave your room without it. When you do leave, turn a light on. Some recommend you put the "do not disturb" sign out to give the appearance the room is occupied. Put your purchases and other items of value out of sight in your room. Most hotel employees are honest, but there is no point in placing temptation within reach.

Don't enter your room if there are suspicious people in the hallway. Walk back to the elevator, go to the hotel desk, and report your suspicions to the management. When you enter your room, lock the door and check to see that doors to adjoining rooms are locked from your side. Do not open your door for people you do not know or services you have not requested. Report these people to hotel management.

When I was staying in a major, well-known Toronto hotel, there was a knock on my door about midnight. I was awake, reading, and watching television, and went to the door to find a young lady who asked for a person with a different name than mine. I told her I was not that person, but she suggested that since she was already there, I might want some company. I politely declined. Had I accepted, would I have been entertained, or would I have been relieved of my belongings? I don't know, but for me the risk wasn't worth taking. You shouldn't take it either.

At night in a strange hotel room, I find it useful to leave the bathroom light on and the door ajar. Despite the need for energy conservation, safety comes first. You may be able to walk around your home at night without stumbling, but you won't be as familiar with your hotel room.

It's helpful to pack a small flashlight in your luggage to be used in the event of power failure in the hotel. If you are taking medicine, be sure to travel with an extra prescription as well as a supply of your medication. If the medication is lost or stolen, you may be able to have the prescription refilled locally. If you wear eyeglasses, carry both your prescription and an extra pair of glasses with you.

If your travels take you to overseas destinations, understand that Yankees are not universally loved. Americans who travel abroad visit, in order of frequency: the United Kingdom; Japan, Germany; the Bahamas; and France. Three of these countries are located in Western Europe, where nearly a third of the world's terrorist attacks in the last thirteen years have taken place. Of those incidents, 40 percent were directed toward United States citizens or property.

Although your chances of being hijacked are statistically low, you might avoid traveling on those airlines that have a history of terrorist attacks. All in all, your chances of being the target of any kind of terrorist attack are probably about equal to your chances of winning $1 million the first time you buy a lottery ticket. But when you travel abroad, put your common sense in gear. Check with the Bureau of Consular Affairs, which maintains a list of nations where travel is not advised. When in another country, let the U.S. Consulate or Mission know you are there. Make sure your passport is valid, record its number elsewhere in your belongings, and carry a copy of your birth certificate.

Keep your profile low, avoid flashing expensive items and looking "important," and stick with the better class of tourist accommodations. Stay on the beaten track; leave excursions into the seamier parts of a town to the natives. And remember to obey all the laws of the country you are visiting. This applies particularly to laws governing the possession and use of drugs. The typical relaxed attitude of our criminal justice system toward drug use is not mirrored in most other countries, and some Americans are sitting out lengthy jail terms for doing something for which their wrists would be slapped in this country.

If I've made travel sound like less fun than it should be, I apologize. I've done a lot of traveling, and while I've experienced some unusual events, I've not yet been robbed, burglarized, or assaulted. Maybe that's because I follow many of these rules, or maybe not. Many people travel extensively and never have a problem. I think that's because they use common sense and travel defensively.

9 If a Man Answers

When I entered the police business years ago, it was a condition of employment that police officers have telephones in their homes. We had frequent debates among ourselves, and occasionally with the Internal Revenue Service, about whether the cost of the telephone was a tax deductible work expense. Our rationale was that we would not have had telephones if we had not been required to have them.

Today, only the unfortunate or the eccentric do not have a telephone. For many, it is a lifeline between the home and emergency services and a link between our place of business and the family. A blessing most of the time, the telephone can also be a downright nuisance, as evidenced by those calls we always seem to get at dinner time from those who would sell us everything from solar systems to storm windows. But more than being a nuisance, a seemingly innocent telephone inquiry, improperly handled on your end of the line, may assist a would-be burglar, rapist, or extortionist in victimizing you. Here are some tips on the safe use of the telephone.

For privacy, you may opt for an unlisted telephone number. An unlisted number will prevent someone from locating your name and address in the telephone book or from getting it through directory assistance. It will also prevent your out-of-state friends, occasional callers, and others who may need to contact you in an emergency from finding you. If you conclude that an unlisted number is not necessary (and for most of us it is not), your next option is to list only your last name and initials, but not your address, in the directory.

That is an advisable option. Most people you will want to welcome to your home will already know your address—or you can provide it when they call. Requiring them to call first will allow you to have that afternoon tête-à-tête with less chance of interruption. Leaving your address out of the book also may prevent a burglar from casing your home via the telephone, a technique I will explain later. Using only your initials will not give away your gender.

Listed or not, don't give your telephone number to every Tom, Dick, and Jane. At least put strangers to the trouble of finding it. It is your private domain and only those you trust with it should be given it.

When the telephone rings, answer with "hello," not your name. While it may seem business-like to answer "Smith residence" or "Susie speaking," unless the person on the other end of the line is someone with whom you truly wish to speak (and you won't know that when you answer) there is no reason to tell him or her who you are. If you do conduct business from your home, I suggest you have separate telephone numbers for home and business and that you turn the business number off, use an answering machine, or engage an answering service after business hours.

If the person on the other end of the line asks "Who is this?" respond by asking "Who were you calling?" While this may seem to be a game of twenty questions, you are justly protecting both your privacy and your safety by being circumspect in your answers. You do not know who is on the other end of a telephone line, where they are calling from, or their motive in calling you.

If the caller asks for someone who does not live at your residence, politely tell him or her that there is no one by that name at your number. If the caller asks what number he has reached, ask what number he is calling. If he is calling your number, reaffirm that there is no one at your home by that name. If he is calling a different number, but misdialed, tell him that he has reached the wrong number, but do not give him your number or tell him your name.

When should you hang up? When the conversation between you and the stranger becomes less than businesslike or when the caller will not answer your questions but persists in pursuing his own line of questioning.

If the caller asks for someone who lives in your home but is not at home, indicate that the subject of the call *is not available* at the moment, but that you will be happy to take a message and have him or her call back. Be especially careful to train children to do this and not to give out information about where parents are or when they plan to return. In the wrong hands, such information makes those who are at home vulnerable if the caller's intentions are less than honorable.

The point of this ''sparring'' with the caller is to refrain from giving a stranger any more information than he may already have, to avoid confirming any information the caller may have, and to avoid letting the caller know who is and is not at home or when they might return.

What about nuisance calls—when you answer the telephone and someone on the other end of the line refuses to speak, or all you hear is heavy breathing? Hang up! If calls like that persist, notify the telephone company and the police. Change your telephone number. You may wish to reconsider the unlisted number.

How about the obscene caller? Unless you're not terribly concerned about your personal safety (in which case you've wasted your money on this book), hang up. There are, from time to time, various tips published for those who receive obscene calls. I must confess that I have never received such a call. If I did, and the caller were female (that being my proclivity), I might be inclined to listen. But most obscene calls are received by women from men, not vice versa, and most women are disturbed by them. My advice is to get rid of the caller as fast as possible, and the best way is to break the connection. If you were making an obscene phone call, would you waste your time on someone who wouldn't listen to what you had to say? You can, as some have suggested, blow a whistle or a loud horn of the type small boaters use in the caller's ear. That may discourage him from calling again, but I don't think it's worth the effort. If you won't listen, he'll most likely not call again.

You certainly should notify the telephone company and the police if you receive an obscene call, but the chances are, unless you receive such calls on a regular basis, little can or will be done about it. It is extremely difficult, despite today's technology and what you may have seen in television drama, to trace random telephone calls, let alone find the caller and prove who committed the offense. Your goal should be to stop the activity directed at you. Cooperate with the police and the telephone company in their investigation of your calls, but the investigative effort rests with them, not you. So don't try your own recording of calls, which may be illegal in your state, unless the police or telephone company advise you to do that. In that case, they will install the necessary equipment and teach you how to use it.

What are the chances of the obscene caller, the heavy breather, or the silent party on the other end of the line arriving on your doorstep? Not great. After all, people who choose this method of communicating do so because they do not desire or can not tolerate a face-to-face meeting. So, generally speaking, while such calls can be alarming and are a nuisance, you should not expect to encounter the caller.

What you may expect to encounter, however, is a person who "cases" your home by telephone before attempting to break in. If he has already selected your home or the general area in which you live as a target, he can find out if you are home by calling you. If you are at home, he probably won't be over to see you, unless he has something more in mind than burglary. If he is a robber or rapist, he may want to find you home alone. That's why you shouldn't give any information over the telephone. If he calls when no one is home there's not much you can do about that, unless you install a telephone answering machine which says, in your voice, "Hello, I am sorry that I'm not available to take your call. . . ." You haven't told him you're not home, and you haven't told him you are. He won't know when the home is empty unless he physically cases it. If he does call your number and gets no answer, at least he's forced to guess about when you might arrive home.

Cost and convenience will dictate how many telephones you have in your home and where they are located, but you should have one adjacent to your bed, with emergency telephone numbers clearly marked upon it. In an emergency, if you don't have the right numbers or can't see them, don't panic. Call the operator. That will take longer than dialing directly, but it will get the message through.

Be sure you know which emergency agencies serve you—in which jurisdictions you reside. Calling the wrong one will delay response to your home. When you reach the emergency dispatcher, tell him or her the exact nature of your emergency. Give the dispatcher your telephone number so that, if you are disconnected, you can be called back. Give your complete address and be sure to state whether you live on a court, street, or lane, then give your name. Stay on the line until you are told by emergency personnel to hang up.

If you are disabled, there are devices you can wear that will seize your telephone line and dial an answering service or an emergency agency when you press a button. A recorded message on your end of the line will tell the answerer that you need help and direct assistance to your home.

In many communities, calling 911 will summon assistance, but be aware that 911 service is not available everywhere. Where it is, it usually is faster to dial 911 than to attempt to dial the specific emergency agency directly. In many areas you can dial 911 from a pay telephone without depositing money. Other features of some 911 systems include the ability of the operator to see the telephone number and address you are calling from at the time the call is received.

The telephone is a favorite tool of the extortionist. If you are perceived as being wealthier than the average citizen, or are in an occupation where cash is readily available, you or a member of your family may become the target of such an opportunist.

Say an extortionist calls to demand money. He or she may tell you that a loved one has been kidnapped and that if you don't pay, and pay quickly, something dreadful will happen. You may believe the extortionist, because he sounds serious and is convincing. So you call home to your wife, Bertha, but there is no answer. Since Bertha is always at home this time of day, you know the caller wasn't kidding.

But it's a hoax. Extortionists depend upon your believing that the caller has kidnapped someone. It's a lot easier (and less risky to the extortionist) to convince you that someone has been kidnapped than to actually kidnap someone. If he or she is caught, the penalties are not as great. So, before the extortionist calls you, he will call Bertha. He may tell her that on your way to work, you were seriously injured in an automobile accident and were taken to Mount Samantha Hospital. You are in critical condition, he will say, and Bertha should come right down to the hospital. To make it more convincing, he will tell her that he is Officer Jones of the local police department. Bertha, a kind and loving mate, runs out to the car and heads for the hospital.

If Bertha had stopped to call you to see if you were all right, or had called the hospital to confirm the report she had received, or had called the police department to verify Officer Jones' report, she would have found that something was amiss. And, if you call your next-door neighbor to see when and how Bertha left the house, you might find that she left by herself in a hurry. She might even have told that neighbor where she was going and why.

When Bertha gets to the hospital, it will take her some time to find that you're not there. In the meantime, you may have already deposited the money the extortionist demanded in the garbage can down the street, since he told you to get it there within thirty minutes or you would never see Bertha again.

There are many variations on this theme. To give himself more time, the extortionist may disconnect the telephone line at your home. You will call and get no answer. Or he may call Bertha, identify himself as a telephone company employee, and ask her to leave the telephone off the hook for testing purposes. Since extortion is not the major theme of this book, I won't go into all of the ways in which it is accomplished. My point to you is that the telephone can be a weapon in the hands of an extortionist. But, it can also be a powerful weapon to you and Bertha in fighting back, if you use it to try to verify information you have been given, to verify the identity and authenticity of your callers, and to verify each others' whereabouts.

That is not to suggest that you should keep such calls to yourself or that you should attempt to handle them alone. Always call the police, don't leave until they arrive, and don't take any money anywhere unless the police advise you to do so.

When Alexander Graham Bell first dreamed of his invention, little did he dream of all the ways in which it might ultimately be used.

10 Defense Mechanisms

Suppose you've done everything I've suggested thus far, but you're still not comfortable. You want some armament, some kind of weapon or weapons with which to protect yourself. There are many individuals and companies today that will encourage you, even teach you, to arm yourself. As a backlash to the issue of gun control, some towns have even passed laws requiring you to own a weapon. I'm not in favor of citizens arming themselves, but since you may elect to do so anyway, I've included some words of advice.

Unless you are aggressive by nature and are emotionally and physically prepared to injure or kill to protect yourself, you may come out second best in an encounter with a mugger, rapist, or burglar even if you are armed. In fact, you may be in a worse situation if you are armed than if you are not. Why? If the criminal is not armed and you are, he may take your weapon from you and use it against you. If your assailant *is* armed, he may take more drastic action than he might have taken had he found you unarmed.

If you insist on getting a weapon, a gun is both the best and the worst. A gun is the most deadly, least personal, and easiest to use weapon there is. The ownership of a handgun is illegal in some places, but in most places you are entitled to have a gun in your home or place of business for the purpose of self-protection. That does not mean that you can legally carry it back and forth. Carrying a concealed gun is probably illegal in your jurisdiction without a special permit, which you may or may not be able to get. Local laws may allow you to carry a gun provided it is not concealed. You'll have to check the laws in your community. Assuming that it is legal for you to have a handgun or to carry it, should you do either?

A high percentage of the persons maimed or murdered in the United States are done in by someone they know personally. By owning or carrying a weapon, many of these victims may have provided a weapon to someone who then used it against them.

A criminal is not the only threat to your safety when you have a gun around. For example, in a recent year, private citizens in California killed fifty-eight felons who were either committing or escaping from crimes. That's good news, until you discover that, in the same year, ninety-nine Californians accidentally shot themselves to death.

A study conducted by James Wright and Peter Rossi of the Social and Demographic Institute of the University of Massachusetts reveals that about 50 percent of American families own a firearm and that number has been fairly constant since 1959. About 7 percent of the adults who own handguns carry them for protection outside their home, and between 2 and 6 percent of those who own guns have fired them in self-defense.

Most people who own guns use them for sport or recreation; about 25 percent use them for self-defense. About 250,000 guns are stolen each year, and 30,000 deaths—accidental, homicidal, and suicidal—are attributed to guns annually. In contrast, there were 150 people killed with guns in the combined countries of Great Britain, Japan, Canada, and West Germany in 1979.

By providing a weapon at the crime scene, wherever that may be, the person who intends to use it to protect himself or herself places it within reach of the criminal. The question then becomes, who is in the best position to use it? If your assailant has the drop on you, even if you have a weapon, the element of surprise you would want to work in your favor is on the side of your assailant instead. You are in a position of reacting, rather than initiating action, and that is the second best position. Can you find your weapon, draw it, and use it before your assailant or intruder can find it or take it away from you? In the event your intruder is already armed, you have a stand-off. Is there more incentive for you to use your weapon or for your assailant to use his? Your assailant is probably not new in this territory—you most likely are. Who do you think has the best chance under these conditions?

If you decide to own a handgun, buy a good one. No Saturday night specials! Go to a reputable gun dealer and buy a weapon made by a well-known manufacturer. If the gun is used, make sure it is in good working order. Buy a holster in which to keep it and a box of ammunition. Go directly to an organized, certified firearms organization, and sign up for professional instruction. In target practice, shoot the kind of bullets you are going to carry in your gun, not the ''wad-cutters'' or plastic bullets sometimes used in target practice. Practice, over and over again, until you are comfortable with the weapon and can hit the bulls eye—the broad side of the barn is not good enough. Practice both single and double action firing, terms I'll explain in a moment. Repeat this practice at least once every six months, the more frequently the better.

There are three basic types of handguns: those that fire a single shot; revolvers; and automatics. The limitations of a single shot weapon should be obvious. You have one opportunity to hit your target before you must reload for a second shot. Chances are good that you won't have time to reload in an emergency situation.

A revolver has a cylinder that usually holds five or six bullets (also called shells or cartridges) and that rotates from one shell to the next as the gun is fired. It may be fired by cocking, or pulling back, the hammer, and then pulling the trigger (called single action) or by pulling the trigger without first cocking it (called double action). Firing double action requires more physical strength and is more difficult to do accurately than firing single action.

An automatic requires three motions to fire. First, you remove the safety, usually by sliding a lever similar to a switch; second, you pull back the breach, which puts a shell into the chamber; finally, you aim and pull the trigger. An automatic must be maintained more carefully than a revolver and is more temperamental to use.

My bias is against automatic pistols, but many will disagree with me. I believe a revolver is safer and less complicated, particularly for the novice, than an automatic. The term automatic, by the way, is a misnomer. The weapon is really semiautomatic. It fires a shell each time you pull the trigger. A true automatic, illegal to own without a federal permit, fires a number of shells in rapid succession when the trigger is pulled and held back.

Of 105 law enforcement officers killed with their own handguns in the period 1973 through 1979, all were killed with revolvers. But, one argument is that since it takes significantly longer to fire a semiautomatic pistol than it does to fire a revolver, those officers might have survived had they been carrying semiautomatic weapons.

In a test reported in *Police Chief* magazine, it took law enforcement officers less than a second to fire a revolver, compared to almost seven seconds to fire a semiautomatic. In a test of civilians, it took nearly two seconds to fire a revolver, compared to more than sixteen seconds to fire a semiautomatic. Of course actual times vary, depending upon the familiarity of the persons being tested with the weapons they are using. The study emphasizes the difficulty and slowness of firing a semiautomatic weapon and hypothesizes that because someone who took a weapon from a police officer would have difficulty firing it, the officer could either regain it or escape. However, one could argue that because it takes so long to fire a semiautomatic, more, rather than fewer, police officers might be killed, because they would be hampered in returning fire when fired upon. The same could hold true for you and me if we depended upon a semiautomatic for our protection.

If you do carry a gun, carry it on your person, not in your car or some other place, but make sure you can do so legally. It's a little simpler for men to conceal a gun on their person than for women. All those cute places you see in the movies, like in the bra or in a tiny holster strapped to a thigh, are more titillating than practical. Women most often carry their weapon in their purses, frequently underneath everything else. That makes it difficult to find in an emergency. If the purse is grabbed, chalk one up for the purse snatcher who, by virtue of your generosity, is now an armed robber. A man has the option of either shoulder or belt holsters, and a boot is not a bad place to carry a weapon, whether you're male or female, provided you have the right holster.

When should you use a weapon? Only when you are prepared to kill and when the consequences—killing another person—are sanctioned by law because of the seriousness of the situation. You do not want to spend time in prison for overzealously defending yourself from a crime that society believes is not serious enough to warrant the taking of a human life. My rule is that you should not use your weapon unless your life, or the life of another innocent party, is in mortal danger, and all other means to escape the situation have failed. That means that you do not shoot somebody to deter having him or her steal your property, and you do not shoot someone who is escaping from the scene of a crime. This is a highly complex philosophical, ethical, and legal issue that you must decide for yourself. I can tell you that killing is being sanctioned less and less in such situations.

Why do I talk of shooting to kill, rather than to wound? Because when you shoot in defense of yourself you shoot to stop the other person. You aim for the largest portion of your attacker. You cannot be sure that your shot will not kill. Therefore, you must be prepared for that possibility, both emotionally and legally.

Should you draw your weapon against someone who already has one pointed at you? Not unless you always win the company baseball pool and come out ahead each time you visit a gambling casino. If your assailant has the drop on you, politely, even graciously, give him or her what he or she wants, unless that assailant clearly means to do you in permanently. Then, what have you to lose? If it's your belongings he wants—let him have them. If it's your virtue he wants, and he is armed, give him that too. You can't replace it, but you can live without it, despite what you may have been told.

If someone is pointing a gun at you, remain as calm as possible. Watch the gun, continue to talk to its bearer, and slowly twist your hips to the right and left. If your assailant does shoot, that motion may take you just out of the line of fire. Comply with what the gunman says, but try to keep both of you calm. If he does shoot, fall to the floor or ground and lie very still. It's called "playing possum."

What if you don't want to carry a weapon, but want one at home for protection? You reduce your exposure to shooting yourself, but you increase the likelihood that persons close to you will shoot themselves or you. I could both regale and sadden you with tales of experienced police officers, trained in the use of weapons, who shot themselves in the buttocks while sitting on the john, or who blew holes in patrol cars roofs with shotguns, or who shot their partners while showing their weapons off in the patrol car. I can recall one of my staff who blew a hole in his desk while cleaning his semiautomatic, and the son of one of my former police associates, who, with Daddy's gun, carelessly left on the bed at lunchtime, "blew away" Daddy's flashlight, also left on the bed.

Then there was the member of my own family whose grandson found grandpa's gun in the nightstand drawer, and who pointed it, for what seemed an interminable time, at grandma, until he was talked into putting it down. He wasn't mad at her. He was just playing cowboy. There is also the case of the husband of a former employee who shot himself in the knee while practicing his sport, which happened to be quickdraw shooting. None of these specific situations was fatal, but many others were.

In one police department in which I served, three different police officers were accidentally, but just as fatally, shot by other police officers while handling police calls. I could also relate stories of police officers I have known who used their weapons to end their own lives. Outside the police department, I can tell you of any number of cases where a wife shot her husband as he was coming home late after having imbibed a few with the boys, and reverse cases exist as well. The surviving party inevitably states that he or she thought the other was an intruder.

In separate incidents in neighboring communities recently, a ten-year-old and a thirteen-year-old fatally shot themselves with their father's weapons. In each case they left notes indicating trouble adjusting in school.

On a more personal note, I can relive the night when I awoke with a start to see someone crawling through our open bedroom window. I lay there very still, waiting for my breathing to become more regular, my night vision to improve, and to reach for my gun, when I realized fortunately, that it was not a prowler crawling through the window but my wife looking out the window.

Of course, to avoid some of these hazards, you can keep your gun under lock and key. But what good will it be when you need it in a hurry? When my children were very small, I installed a lock on the top dresser drawer and stored my service revolver there. If I had small children I would do the same thing today, but the weapon was inaccessible in the event I needed it in a hurry in the middle of the night.

Gun safety rules for range firing or hunting won't always apply in self-defense situations. If you are going to have a firearm there are some basic rules you should follow. Always treat every gun as though it were loaded. Never point a gun at anything you don't intend to shoot. Never lay a weapon down where someone else may pick it up. Practice with your weapon and ammunition frequently. Keep your weapon clean and well-maintained.

What about a weapon you carry unloaded, just to scare the other person? That person won't know if it's loaded or not, will he? Forget it! That person will assume it is loaded, and if that person has his or her own weapon, you may bite the dust. Don't ever bluff. Never carry an unloaded weapon.

If a gun is not for you, what about a knife? Knives are much more difficult to use than guns. Stabbing or slashing someone is a much more personal issue and requires more strength (and a much stronger stomach) than pulling the trigger on a gun. There aren't many classes on how to use a knife to defend yourself. Unless you grew up on the streets and learned to use a knife as a matter of self-preservation, you will fare poorly with one. On the other hand, your assailant will probably be much more familiar with the handling of a knife than will you.

Today, encouraging you to carry and use tear gas is in fashion. The people who benefit most from this fad are the tear gas manufacturers and the instructors who are paid to teach classes and sell you the tear gas. Tear gas will work as a repellent on many people, but it won't work on all. Those who are mentally crazed or who are under the influence of chemical substances may not be affected by it at all. Further, armed with tear gas and the platitudes of those who champion its virtues, you may develop a false sense of security and go places and do things you wouldn't otherwise venture.

Like almost everything else, tear gas has a shelf life, as does the material that ejects the gas from the cannister. You must test it occasionally to see if it still works. When you need it, you must make sure you are pointing in the right direction, since it will work against you just as well as it will against an assailant, and the wind must be blowing in the right direction for its use. Picture yourself asking your assailant to turn around and face the windward direction, so that he gets the effect of the gas instead of you.

The manufacturers of tear gas know its limitations. The instructions on one tear gas cannister state, for example:

> The contents are dangerous, use with care. Keep out of reach of children. Contains approximately 60 two-second bursts. This unit forcefully projects a stream about ten feet. The stream must contact the face of the subject, but do not spray directly into the subject's eyes. Contents of can should only be discharged in an upright position.

As you can no doubt tell, I am not a great fan of the use of tear gas for protection by the average person. Nevertheless, many of you will decide to carry it for protection. To use tear gas, you have first to be able to find it. Typically, a person who has just purchased tear gas will carry the cannister in his or her hand when walking. After a period of time, when nothing untoward has occurred, the tear gas will go into a pocket or purse. Later, it will find its way to the bottom of the purse, or it may be left atop the dresser, to be carried only when you know you are venturing into dangerous territory. If you are going to carry it, carry it in your hand where you can use it quickly.

There are those who suggest, in books like this one, a host of weapons one might use. They range from those designed to scare off an attacker to those specifically designed to injure. Weapons to frighten an attacker include such items as a police whistle or any number of other small devices that emit noise. There are gas powered horns that make a piercing sound when a button is pushed, and electric alarms, powered by small batteries, that do the same. There is an attache case with a built-in siren that activates if the case is grabbed. These noisemakers are generally legal to carry around, so long as you don't practice with them in front of the church on a Sunday morning. If they do not frighten off an attacker, they will at least sound the alarm.

Many of the weapons that are often suggested are illegal, if not to own, at least to carry. They include a billy club or nightstick, blackjack or sap, brass knuckles, and similar weapons. Some suggest carrying an electric cattle prod or an even less civilized version designed for use against humans. When these devices were used by a few police departments in the 1960s, they drew widespread public outrage, and there is little sympathy today for the use of such weapons on humans. A gun that shoots barbed darts attached to thin wires that carry a high-voltage electrical charge is generally restricted to law enforcement use and is not widely used in any event. A device designed to blind an attacker by triggering the equivalent of 250 flashbulbs in his eyes is probably legal, but it is certainly inconvenient to carry. Perhaps the most novel weapon is a three-foot alligator carried by a Buffalo, New York, resident. Some items that you wear or carry with you each day can be used as weapons. They include spike heels, now back in style, which work well on the instep or shins of an attacker, as do heavy boots or shoes. If you stomp down with the heel of your shoe on an assailant's foot, you will cause pain and may break bones. But kick the high heels off if you decide to run for it. A fingernail file, rat-tail comb, ball point pen, sharp pencil, or key can be used to distract an assailant if not repel him. Stuck in the assailant's eyes, chest cavity, solar plexus, windpipe, ears, or nose, they may disable him, if you have the strength and stomach for that kind of move. The teeth of a comb can be raked across the bridge of an assailant's nose, and as long as tennis remains popular, you can carry a steel racket with which to whop an attacker. Hair spray aimed at your opponent's face may blind him.

There are also the weapons with which you were born. You can kick with considerably more force than you can hit or slap; aim at the testicles, kneecaps, or insteps. You also have great strength in your jaws. In the San Francisco area recently, a rape victim bit off a portion of her attacker's tongue. He was arrested after he went to a hospital emergency room for treatment. Any bony part of your body can be used as a weapon—your knees, elbows, chin, or forehead.

If your life is in danger—if you are really going to be maimed or killed if you don't get away or do something—anything is better than nothing. It is important, in this situation, that you quickly adjust your mental attitude to that of becoming the aggressor, rather than the defender. If you do not, you are unlikely to be able to use the weapons at your command.

For the truly paranoid, there's the briefcase that has everything but the kitchen sink. It is lined with a bullet resistant fabric that allows you to use it as a shield. Inside there's a six-hour recorder, a miniature voice stress analyzer that helps detect when someone is lying, a portable communications system, a telephone message scrambler, a bomb detector, a detector that tells you when someone is recording your conversation, and a siren alarm that signals when the briefcase is being stolen. Do you need one of these? Only if gadgetry is your bent.

How about wearing a bulletproof vest? They seem to be popular these days, particularly among police officers who have a legitimate need for them. But, if you need to wear a bulletproof vest, you probably need more help than this book can give you. But for those who need them, there is a clothing line that's doing well with its bulletproof fashion collection. The jackets and vest are casual on the outside, and have a lightweight, ballistic-resistant panel on the inside. One manufacturer plans to offer a business suit next; a bulletproof jacket and vest with matching pants.

What about lessons in self-defense techniques like judo, karate, or kung-fu? I recommend them, particularly for children living in urban areas. But unless you constantly practice the skills you learn in these classes and keep yourself in good physical condition, the skills will be of little value to you within weeks of having taken the lessons. Many books that give blow-by-blow descriptions of various self-defense techniques and holds are commercially available. But, unless you are taught these techniques properly and practice them regularly, you will not be able to use them when you need them.

The best weapon? That's the subject of the next chapter.

11 Strategic Weapons

Don't look back. Something may be gaining on you.

Satchel Paige

If you decide to read only one chapter in this book, make it this one. There exists a weapon in the fight against crime that is more powerful than any other. It is not a gun, knife, or club, nor is it kung-fu, judo, or karate. It is not a weapon that can be acquired from an alarm company or a hardware store. It is intelligence.

All of us have brains, but not all of us use them well when attempting to prevent bad things from happening to us. One need not have a high IQ to avoid trouble. What you need is a combination of alertness and common sense.

I once wrote an article wherein I commented that "survival of the fittest" was a phrase commonly used to describe life in the animal kingdom but is now frequently used to describe our difficulty surviving in a rapidly changing environment. One facet of that environment is the increase in crime I talked about in the introduction to this book. In this context, survival means learning to avoid being a crime victim, not so much by defending yourself as by preventing things from occurring. That's what this chapter is all about—using your intelligence and senses to avoid becoming a crime victim.

Crime is an everyday occurrence. All of us are personally affected by it. The most important thing you can do to avoid crime is to learn to be aware of who and what are around you. Develop a healthy alertness, but not an unhealthy paranoia, about what's going on. If you were taking a stroll through a jungle, you would avoid stepping in quicksand, and if you could, you should try to avoid equally unpleasant experiences in a city. Yet, I marvel at the number of people who, after waiting for the pedestrian light to change in their favor, step off the curb without first looking to see if traffic is approaching. They walk smack into trouble without ever seeing it in advance.

Who are the typical victims of crime? Males are the victims of violent crime much more frequently than are females. Men are robbed or assaulted twice as often as women. Young people between the ages of twelve and twenty-four bear the greatest risk of being a victim of violent crime, while, surprisingly, chances are lowest for those sixty-five and older.

If you are black, your chances of being robbed are nearly two and a half times higher than if you are white. Black men have the highest violent crime victimization rate, followed by white men, black women, and white women.

Divorced or separated persons suffer a higher violent crime rate than those who have not been married at all, followed by those who are married, and then by the widowed. If you live in a household headed by a man to whom you are not related, you are in the highest risk group for violent crimes. Men living alone have the second highest violent crime rate. If you are a woman married to the man with whom you live, you are in the lowest victimization group for violent crime.

People twenty-five and older who have attended college for one to three years are more likely to be the victim of a violent crime than college graduates or those with no college education, and, the lower your income, the higher your risk of violent crime.

If you are unemployed you are twice as likely to be the victim of a violent crime than if you are employed. But homemakers and the retired, who statistically are not considered part of the work force, have a lower violent crime rate than the employed. If you are a nonfarm laborer, a service worker, or are in the armed forces, you are more likely to be the victim of a violent crime than if you are in any other occupational group. Least likely to be victims of violent crime are farm owners and managers, and professional and technical workers.

Strangers commit more than half the crimes of violence, and men are more likely than women to be the victims of a stranger. Among females, the elderly are more likely to be victimized by strangers, while whites are more likely to be the victims of strangers than blacks.

The majority of violent crimes are committed by males. Young persons are most often victimized by young criminals, while those over twenty are more likely to be victimized by adults.

More than half of all violent crimes occur at night, and most, except for crimes of rape, do not occur in the victim's home. Rape occurs in the home a third of the time. Robberies occur most often in streets, parks, and parking lots. A weapon—firearm, knife, club, brick, or bottle—is displayed in one-third of violent crimes, and victims are physically injured in 30 percent of personal robberies and assaults.

The profiles of theft victims are somewhat different. Households headed by young people are most vulnerable to burglary and auto theft. In fact, young people are almost five times more likely to be burglarized than are senior citizens. Households headed by blacks or Hispanics are more likely to be burglarized than those headed by whites, and blacks and Hispanics suffer more motor vehicle thefts than other groups.

People with higher educational levels are more vulnerable to crimes of theft, and, the higher your income, the more likely you are to have something stolen from you. On the other hand, families with the lowest incomes experience more thefts than those of middle incomes. The larger the family, the larger the household theft rate. People who live alone have the lowest burglary rates, and if you're serving your country in the armed forces, you will contribute more to the thieves of our nation than any other occupational group.

That's all very interesting, you say, but what can you do about it? One of the most important steps you can take to avoid crime is to train yourself to be aware of the activities around you. When you learn to be observant of individuals around you, you will more quickly spot suspicious persons and activities and be able to avoid them or signal their presence to someone else. Many people who have not trained themselves to be observant see only the landscape, the big picture, when they look at something. They miss the detail, the individual activity that completes the picture.

Test yourself to see how aware you are of what's going on around you. Observe a crowded street corner, sporting event, or theatre. Do you see the crowd, or do you see individual activity within it—people talking with one another, a dog, a blind man, transactions taking place? It is that ability to see individual activity within a group or crowd that's important.

In police work, "follow your hunches" is an often heard axiom, particularly for new recruits. Following hunches pays off. I recall, when I was a police officer, being miles away from the scene of a night-time robbery of a restaurant. Other police officers had responded to the call and were searching for the robber, and my partner and I were not needed there. But we had a "feeling" about it, so we went anyway. We found the robber, the gun, and the money under a car in a parking lot that had already been searched by other officers.

Another time, the scenario was similar. We were miles away, weren't needed, and there were plenty of officers already at the scene of an accident involving a stolen car. We went anyhow and apprehended the auto thief as he nonchalantly walked up the street a block from the accident.

If you have a feeling about something, act upon that feeling. Assume that your feeling is valid and adopt a conservative response to it. If you feel unsafe doing something, don't do it. You need not have an intellectual reason for your feelings. If you are uncomfortable or uneasy about your surroundings or the people you are with, change your environment or the company you are keeping. Do not ignore those subliminal flashes your brain both generates and receives—act upon them.

You are most vulnerable when you are walking in public. Wherever you walk, there are certain tips that can be helpful. First, if you aren't familiar with the area in which you are going to walk, check it out in advance. Talk to people who know the area, look at a map, drive the route in a car or walk it with others, but gather some information about the territory first. Choose well-traveled and, if walking at night, well-lighted streets.

Try not to walk alone, especially if circumstances require you to walk along dark, deserted streets. If you can delay your walk until you can walk with a group of people you know, do so. There is safety in numbers.

When walking along a street, especially at night, walk close to the curb, avoiding dark doorways and alley entrances, and stay clear of shrubbery and parked cars. When walking on the sidewalk, walk facing traffic whenever possible. That allows you to see vehicles and their occupants as they approach you. If a vehicle should stop, and its occupants try to harass or molest you, you will be moving in the opposite direction of the vehicle, making it more difficult for it to follow you. If you can't follow this advice, and a car appears to be following you, turn or walk in the opposite direction, or cross the street and walk on the other side.

Don't take shortcuts through parks, tunnels, parking lots, or alleys. If conditions on the sidewalk are bad, don't hesitate to walk in the street to avoid trouble.

If you are approached by someone you don't know, don't stop—walk around them. Be cautious about pausing to answer questions or give directions; it's best to keep moving while you do so. Don't stop to give handouts, offer a match, or light a cigarette for a stranger.

Keep your mind and eyes on what you are doing and where you are going. Watch for trouble and stay away from it. If you see people fighting on the street, avoid walking near them. If people are arguing, mind your own business. Avoid being jostled. That's a favorite ploy pickpockets use as a distraction, but it may also mean you're not paying enough attention to what's going on around you. If you can't see what is ahead of you as you are walking, be extremely cautious and be prepared for what *might* be ahead of you.

Look like you know what you are doing and where you are going, and have complete control of yourself and your environment. Walk confidently and directly, at a steady pace, and be alert and aware of your surroundings and the people around you. Don't walk with your hands in your pockets—you will appear less than prepared for danger—and don't carry things in your hands if you can avoid it, unless those things are going to be used as a weapon to defend yourself. If you get into trouble, seek help. If you think you are being followed by another person on foot, walk quickly to where there are people and, if it is night, lights.

The relationship between the way you walk and your chances of becoming a street-crime victim has been the subject of a study by Betty Grayson of Hofstra University and Morris Stein of New York University.

By videotaping pedestrians and showing those tapes to a number of convicted criminals, they developed a scale of assault potential and a profile of potential victims and nonvictims, based upon how people walk.

The potential victim is said to walk with an exaggerated stride, shifts his or her weight laterally, diagonally, or with an up-and-down movement, and moves the arm and leg on the same side of the body simultaneously. He or she moves gesturally, so that the arm or leg movements appear to come from outside rather than inside the body, and lifts the feet while walking, rather than moving them with a fluid, swinging motion.

On the other hand, potential nonvictims walk with a medium stride, shifting their weight in a three-dimensional manner; they walk posturally, move opposite arms and legs together and swing their feet rather than lifting them.

If this sounds a bit technical, it boils down to this. Those who were less likely to become crime victims were coordinated in their body movements and appeared to be more comfortable with their movements and bodies. The potential victims exaggerated their normal movements, and their body parts seemed to work against each other.

So much for walking. How should you dress and what should you carry with you? Dress modestly, not provocatively or flashily. While I don't subscribe to the notion that women are attacked more often while wearing low-cut blouses or slit dresses, the less attention you attract, the less likely you are to be selected as a victim. Don't carry any more money, credit cards, or other valuable items than you need, and avoid displaying flashy, expensive jewelry, cameras, or other accessories.

If you are male, carry your wallet in an inside pocket and not in your back pocket. Some recommend carrying it crosswise in the pocket to make it more difficult for a pickpocket to extract. There is even a wallet available which attaches to your belt like a holster. Women should carry their purses with an arm through the strap and a hand firmly holding the side and bottom of the purse. Don't dangle your purse at your side, and if you won't be needing it, don't carry it at all. If a purse snatcher insists on tugging it away, let it go. The purse and its contents are easier to replace than your arm. As I suggested earlier, if you must carry a large sum of money, consider carrying it in several different pockets. While you'd rather not lose any, better to lose some than all. Don't flash your money.

Mothers used to call it "mad" money and always made sure

their daughters carried it on dates. Whatever you call it, carry enough money for bus or cab fare in the event you need to make a quick getaway or your plans abruptly change. Speaking of change, carry enough of that with you to enable you to telephone for help if necessary, and carry any pertinent medical information which others may need in order to treat you in an emergency.

Don't leave your purse on a counter while you are paying for something. Don't walk around with it open. In your office, lock it inside your desk or a cabinet. Don't leave it in a department store fitting room. Don't hang it on a restroom door hook, and don't leave it on the floor of a restroom stall. Don't leave it next to your chair or hanging on the back of the chair at dinner. In short, hold onto it or at least keep it where you can see it. When sitting, your lap is the best place for your purse. Don't walk out of a restaurant or bank or away from an automated teller machine while you are still putting your money away. Put your money away before you begin your walk.

Automated teller machines are marvelous devices, and you'll be seeing more of them, but there are some commonsense precautions you should take when using one. First, keep your card and your personal identification number to yourself, and don't keep them together. You should memorize your identification number, then write it down in a secure place. Don't write it on the back of your ATM card. If your card is stolen, don't respond to someone who calls on the telephone to say that he is your banker and needs your number in order to cancel your card. Banks don't need your identification number to do that.

Anyone who has both your card and your number has access to the money in your account. If you loan them the card and the number, you have just gone into the money lending business. Report any discrepancies in your account to your bank immediately. Some banks take a photo of each person using an ATM, so if you try to con them by saying you didn't make a withdrawal when you did, be prepared to be confronted with a candid snapshot of you punching the buttons on the machine, with the date and time of the transaction printed on the photo.

All the rules for where to walk and what not to do in strange surroundings apply when you are using the ATM. When you go to the machine, park as close to it as you can. Make sure there are no suspicious characters lurking around. It's best to use the machine when other customers are around. If you don't feel right about the surroundings, make your transaction some other time. Don't stand on the street counting your money. Put it in your pocket or purse and count it when you are in a more secure environment. Save your receipt to show to the bank if there's been an error.

A criminal frequently knows the habits and routines of senior citizens—the days when social security or other checks arrive or when the recipient is likely to go to the bank and return with cash. Social security and pension checks can and should be sent directly to the bank for deposit to your account. That way, you'll also avoid having them stolen from the mail box. Use a checking account or credit cards to pay your obligations, rather than cash. If you have to go to the bank, go with a friend.

Muggers prey where there are people. They have a keen eye for gold chains, platinum watches, diamond rings, and fat wallets. In the words of one mugger, "A lot of people give themselves away by having a lot of jewelry, minks, expensive suits, gold, and diamonds." Muggers in New York City and elsewhere are sophisticated enough that they can tell the value of a chain worn by a passerby with a glance at the clasp. Muggers like shopping malls, hotels, hospitals and prime tourist areas. Don't underestimate them. If you say, "You'll have to shoot me to get my money," a mugger may do just that. Muggers have their own rules. They may not let a victim go into his pockets for his money, because the victim may have a gun. Since muggers understand the value of surprise, many attacks are from behind. There is a theory that muggings are more likely to occur at a point of transfer—where you get off the subway or bus, where you get out of your car, or where you turn a corner.

There are different types of muggers. A professional knows what he is doing, where he is most likely to find the prey he wants, and how to practice his trade. Do what he says—don't argue or joke with him and don't try to get away. He knows his territory better than you do, and he may have an accomplice. He is not impulsive; his activity is planned.

A purse snatcher is more likely to be young and inexperienced—just entering the criminal job market. If you give up your purse when he tugs on it, he will be gone. If you hang onto it, you might discourage him, but more likely he will knock you down to get it. Then there is the junky, pot-head, or drug user by any other name who is stealing to maintain his habit. He is likely to be unpredictable and dangerous, and he may be armed. Even worse is a gang of young people who will attack like a flight of locusts. They may number in the dozens and will steal anything, even strip you of your coat or shoes, beat you, and leave you lying on the sidewalk.

If someone grabs you, make as much noise as possible and get away if you can. If you can't avoid being grabbed and can't get away, don't try to fight off the attacker unless you have mentally and physically prepared yourself in advance to do that. The decision whether or not to resist an attack is a personal one and depends upon the circumstances and your readiness, but my general advice to most people is don't resist. There is not agreement upon that. Some people argue that a sudden, aggressive resistance, like a scream or a slap, will scare off an attacker. One study of sexual assault cases revealed that a woman who angrily and loudly curses her molester is less likely to be raped than a woman who pleads, whimpers, or cries. But others maintain that any show of resistance can jeopardize the life of the victim, for no one knows in advance how an assailant will react or what motivates him. So, I stick with my original advice.

Don't chase a culprit. Call the police and let them do that.

If your attacker is going to beat you no matter what you do, refer to my advice in the previous chapter. If you can't or won't fight, lie on the ground with your knees tucked into your stomach and cover your head with your arms. You will be hurt, but less seriously than you would otherwise be.

Some special guidelines are appropriate for those who use buses and trains. First, check the schedules to limit the amount of time you must spend waiting, especially if you travel at night. While waiting, be alert to what is going on around you. Stay away from troublemakers. Don't be obvious about it, but quickly and quietly move away from a boisterous group or a rowdy drunk. Wait near a ticket booth, a station attendant's area, or a well-used entrance. While waiting for a bus by yourself at night, standing in the shadows may conceal you from passersby, and you can step out when you see the bus approaching. Don't accept a ride from someone you don't know, even if waiting makes you uncomfortable.

If, when you reach your destination, you are uncomfortable about getting off because of the people waiting on the street or because you think you will be followed, don't get off. Go to another stop where you can get a cab or another bus or train to take you back to your destination.

While riding, try not to sit next to the door, where you are fair game to whomever gets on. Thieves have been known to quickly board a bus, grab a purse, and get out the doors just as they close. Sit close to the driver. An aisle seat will make it easier to move if you don't like your seat mate. Don't go to sleep, and do hang onto your purse or briefcase, which is safer in your lap than under or over your seat. Statistically, you are safer while riding mass transit than you are walking on the street, but do not develop a false sense of security because you are riding in a public vehicle.

Some bandits find their prey in the restrooms of America. They will accost you while you are seated in a most vulnerable position, demand your wallet, ring, and watch, and be gone before you can get up. When using a stall in a restroom, most people will use the stall furthest from the entrance, where there are the fewest people. When you are using a public restroom, pretend you're at the theatre and sit up front. Speaking of the theatre, if you go by yourself, avoid dark corners. Sit next to an aisle so you can relocate quickly if you don't like your neighbors.

Another type of bandit is the hotel robber. He is usually armed and often finds his victims in an elevator or just outside one. If you are uncomfortable with the people accompanying you in an elevator, push several floor buttons. This gives you the option to exit on any of several floors and won't tip your companions to which floor is really yours.

There is a lot written about rape today, but, unfortunately, most of it deals with how to deal with your feelings after you have been raped. Rape is usually a violent crime, not one involving the uncontrolled passion of a sexually aroused male. It is an attempt to hurt someone, not a means to satisfy a sexual urge. In over one-third of reported rape cases, the rapist is an acquaintance, neighbor, friend, or relative of the victim. Not a small number of rapes are the result of a meeting in a tavern or bar and the acceptance, from a stranger, of a ride home. Know your companions well before you place yourself in an environment that may lend itself to sexual molestation. Never accept a ride from someone you don't know or know only casually. Rape can happen to anyone of any age, and it subjects the victim to physical peril, psychological stress, guilt, anxiety, and anger. Although infrequent outside of prison walls, men can be victims of rape too.

A study in 1976 by the Queens Bench Foundation found that there was a correlation between avoidance and not being raped. About half the successful and unsuccessful rapists studied had first engaged their victims in conversation. Those potential victims who were not raped were the ones who sensed something was amiss and brushed off their new acquaintance. Those who were raped tended to rationalize or dismiss their feelings of uneasiness. The study also found that rapists often leave an escape opening before they attack. In those circumstances, those who are raped are those who fail to seize the opportunity to escape. Remember what I said about playing your hunches.

Other studies indicate that women who walked firmly and purposefully and who were aware and relaxed were less likely to be attacked. In short, the more self-assured a woman appears to be, the less likely she is to be the victim of a rapist.

It will do no good to protect your home if you open your doors to any and all who knock. Don't let people into your home unless you know them and they have business with you. If someone delivers a package you didn't order, or even one you did, have them leave it on the porch. Don't allow strangers in on the pretext that they need to use your telephone. If you want to help, offer to make a call for them, but leave them on the other side of your locked door. If you've just hired a painter, he won't do you much good if you don't let him in when he arrives. But, when a service person you didn't request shows up, don't let him in without checking with the firm or utility first. And don't just call the number he furnishes; look for the firm's number in the telephone directory and see if the number he offered is the same as that in the book.

Recently, in California, a bank extortionist was hard at work. He alternately posed as a delivery man, a service person, and someone interested in the welfare of animals who needed to use the telephone to report a dog that had been struck by a car. Once inside a home, he would produce a weapon and tie and gag the female occupant who, in each case, was the wife of a bank branch manager. In some cases he would record a message from the wife to add authenticity to the demands for money he made of the managers. Some intended victims successfully avoided this man by simply refusing to open their doors to him.

You obviously should not invite total strangers into your home for any reason. So that the person at the door doesn't know that you are alone as you are answering the bell, you could pretend to be talking to someone as you walk to the door.

Don't gossip with your neighbors and acquaintances about your jewelry, stamp collection, coin collection, or other valuables, and caution your children not to tell their friends at school about dad's collection of antique weapons or mom's assortment of sterling silver. Be careful about what gets into the newspapers. Notices of births, marriages, family reunions, vacations, anniversary celebrations, and funerals tip everyone off that your home will be empty at certain times on certain days. The next time you see your name in print it might be in a story about a burglary that took place while you were gone. If you must "hype" yourself, tell the newspaper what you did *after* you return from your trip.

Always let someone know where you expect to be. If you don't show up, someone will know something is wrong and begin to look for you. If you see or hear something suspicious, call the police. If you think you hear an intruder in your home, call the police. Don't try to apprehend an intruder yourself. In chapter 4, I told you that your best weapons against a burglar are noise and light. So, if you hear something in your house, call the police, make lots of noise, and turn on lots of light. If you should happen to surprise an intruder, cooperate—don't fight.

By not thinking, by not being observant, by not going over in our minds how we would handle a given situation, by doing stupid things, or by going to the wrong places, we are usually the ones who place ourselves within a criminal's grasp. We can keep ourselves out of trouble most of the time by doing the correct things. After all, crime is not a random activity for the criminal. It appears random to us because the victim of each crime is usually someone we don't know and can't identify with. But we know, or should know, the signs of danger, and when those signs appear, if we recognize them and take appropriate action, we can avoid danger.

12 Closing the Barn Door

What do you do when, despite the precautions you have taken, you become a victim? First, stay as calm as possible. Second, don't resist unless it is clear that your life is in danger. And third, do not pursue the offender. While I am a strong advocate of self-help, I do not advocate foolish heroism.

Call the police immediately. If it's the type of crime where the suspect is likely to still be in the vicinity, quickly notifying the police enhances their opportunity to catch the culprit. If you should return home to find the doors unlocked or open or windows broken, call the police and await their arrival before going into the house. If you enter your home and surprise a burglar, turn around and run outside. If you can't get outside, look for an inside retreat, like a bathroom with a sturdy door and a good lock. The burglar won't easily be able to get at you, and chances are he'll leave after being discovered.

Report all crimes to your law enforcement agency. They may or may not respond to take a report or make an investigation. That depends upon their view of the seriousness of the crime and their chances of apprehending the offender. Generally, they will respond to calls of homicide (murder), rape, robbery, aggravated assault, or a burglary in progress—crimes that involve injury, death, or the threat of it, or where the suspect may still be in the neighborhood.

On the other hand, depending upon the size of your community, the day of the week and time of day, and the policies and procedures of your police agency, you can probably count on your report being taken over the telephone, or by mail, if your bicycle is stolen, someone makes off with your garden hose, or your wallet is taken from your coat while you are at work. The fact that a police car, with lights flashing and siren screaming, is not going to pull up in front of your home in response to your call should not discourage you from calling. If your life is in danger, the police may arrive in a noisy fashion. If you call to report a robbery or burglary in progress, they will probably respond quickly but quietly—without sirens and flashing lights. To do otherwise might endanger your life and reduce their chances of catching the culprit. The lights and siren response occurs most often on television.

In many crimes—a wallet theft for example—there is simply no practical reason to dispatch an officer to take a report. There is no physical evidence to gather and no suspect to arrest. The report can be taken just as well over the phone and at less cost to taxpayers.

Why report such a crime at all if that is the case? First, your insurance company may require it. More importantly, those reports allow the police to gather statistics indicating where crimes are occurring, so that they can concentrate their resources where they will do the most good.

When you call the police, know how to give them information which will be of most help to them. They will want to know the same things a newspaper reporter would want to know if you were being interviewed for publication. Who are you? What is happening? Is it happening now? Where did it happen? When did it happen? How did it happen? Is anybody hurt? Is the person who did it still there? Where did you last see him or her? Where are you calling from? Where will you be? Be prepared to quickly give precise answers. The more organized your account to the police, the better and faster they will be able to respond.

If your loss is covered by insurance and you intend to make a claim, be sure to notify the insurance company promptly. They will want information similar to that needed by the police.

When you are a victim, you are also a witness. For example, you may have seen the culprit. You, and perhaps only you, can describe the damage, loss, or injury. How well you recall what happened and your description of how it happened will affect the ability of the police to locate and arrest the suspect or recover your property. The capacity of the criminal justice system to prosecute, convict, and punish the responsible party also depends upon your description of the events leading to, during, and after the crime.

In order to be able to describe events, you must be aware of them. In chapter 11 I gave some tips on how to train yourself to be aware of the activity around you. The test of your powers of observation will come when you need to describe what happened and what you saw.

There are some helpful hints for describing suspicious people or those who have committed a crime. While it's not easy to describe someone you don't know or have seen only once, and then perhaps only fleetingly, it's easier if you start at the top. Describe the person's head and hair. Look at the shape of the head and the type, style, color, and length of hair. Next, describe the forehead, ears face, eyebrows, and eyes. Ears may be described in terms of size and shape; faces may be round, square, or oval; foreheads may or may not be prominent; and eyebrows may be arched, straight, long, short, bushy, or plucked. Eyes are of different shapes and may be clear, bloodshot, watery, red-rimmed, etc. Describe the person's nose, mouth, chin, and neck in a similar manner. While you may not remember all of these features when describing someone, if you look for features you will be able to describe some of them. By following a pattern in your descriptions, you'll become a better observer.

Describing a vehicle is a little easier, and there's even a helpful acronym for the purpose. It's CYMBAL. CYMBAL stands for *Color, year, make, body style,* and *license* number. Make refers to the brand name of the auto—i.e. Ford, Chevrolet, Honda—and not to model names like Cutlass or 280Z. Body style refers to whether the car is a four-door sedan, a hatchback, a convertible, and so forth. The most important part of a vehicle description, because it can connect a name with a vehicle, is the license number. Try to see it, remember it, and write it down as soon as possible. But remember that license plates can be switched from car to car, so, while the license number is most important, any description of a suspect vehicle is useful.

Cooperate in the prosecution of someone who has committed a crime, but be cautious if you are asked to sign a formal criminal complaint against someone. If you are the victim of a felony (more serious crimes, like murder, rape, robbery, and burglary are felonies), it is unlikely that you will be asked to sign a criminal complaint. You may be asked to make a statement to the police or the district or prosecuting attorney and to sign that statement, but that is not to be confused with signing a criminal complaint. Nor is the call you might have made to the police in the first place a criminal complaint.

In less serious situations, or in situations where the evidence boils down to your word against that of the accused, you may have to sign a complaint if the accused is to be prosecuted. This is common in domestic quarrels where some violence has taken place before the police arrived and the parties are disputing what happened. It may also occur if you accuse your neighbor of stealing your garden hose, and there is no independent evidence that your neighbor took the hose. If you want him prosecuted, you must sign the complaint in most jurisdictions.

When you sign a criminal complaint, you become, in effect, the arresting officer, although the police may haul your neighbor away and lock him up as a favor to you. (In some jurisdictions, the police may not refuse your request to haul the suspect away once you have signed a formal complaint.) If you sign the complaint, and the person pleads guilty or is found guilty, all is well for you. But, if he or she is found not guilty, the now aggrieved neighbor may be able to sue you for false arrest or malicious prosecution, and if successful, he may collect more from you than the replacement cost of your garden hose. So assist in a prosecution, prosecute if the circumstances warrant, but be sure of your facts. It would be wise to consult your own attorney before you sign a criminal complaint against anyone.

You may find yourself testifying in court, either as a victim or as a witness. Court appearance can be a fascinating, rewarding experience, but it can also be an emotional, frustrating one. The frustration comes largely from the slowness and inefficiency of court proceedings and from the fact that you may be called upon to give the same testimony several times. You may also find that, having summoned you to testify, the judge will find himself too busy to listen to you on the appointed day, and so he may ask you to return. And return you must, perhaps several times.

Despite these frustrations, our system of justice won't work at all if you and I don't help it along by volunteering what we know when asked to do so. It makes sense to be well prepared when we do that. Having made more court appearances than many lawyers, I offer this advice. First of all, prepare. Review any notes you may have made at the time of the incident, any statements you gave the police, and any reports you furnished to others. If necessary, revisit the scene of the incident in order to refresh your memory about what happened. But remember that if time has passed, the scene may not look exactly the same on the day of your visit as it did when the incident occurred.

When you appear in court, be on time and be properly attired. You are not on trial, but your appearance and demeanor will either add to or detract from your credibility as a witness, and we all want to be believed.

Answer questions politely and courteously. Consider each question carefully before you answer it. If you don't hear or understand a question, ask that it be repeated. Speak in a conversational tone, but loud enough for the judge and jury to hear you. Direct your answers to questions to the person who asked the question, usually one of the lawyers or the judge, and look directly at that person as you answer.

Be truthful in your answers, do not volunteer information, and don't guess or surmise. If you don't know the answer to a question, say so. Unless you qualify as an expert witness in some field, you won't be allowed to offer your opinions, and you probably won't be allowed to testify to what someone else told you. That's considered hearsay evidence.

Although a courtroom can be intimidating, stay as calm as possible, don't argue, and don't allow yourself to be badgered. Remember that it is the responsibility of the judge to assure that you are properly treated as a witness. Give forthright, honest testimony, and don't let your personal feelings about the case creep into your answers. To do so may influence the judge or jury to disregard your testimony completely. When your testimony is completed, leave the courtroom with a sense of having completed your civic responsibility.

If your property has been stolen, you will have to provide the police with an accurate description of it. You should have a list of all items you own that have serial numbers or other identifying characteristics. As I mentioned earlier, police organizations and community groups encourage the use of engraving tools to carve your initials, driver's license number, or social security number on valuable items. Without such identifying marks, your stereo system, which looks exactly like a lot of other people's stereo systems, may be sold at a police auction if recovered. Although you (and maybe even the police) may be convinced that an item was taken from you, if it cannot be uniquely identified, chances are good you won't get it back.

If credit cards, checks, or keys are stolen, you need to take steps to minimize future losses due to the use of those items by somebody else. You should have a list of all credit cards issued to you, with complete numbers, expiration dates, and company names, and you should also list the telephone numbers you must call in order to report them lost or stolen. Once you report a loss to the credit card company, you are not usually liable for any charges made with the card, provided it really has been lost or stolen and was not borrowed by your Uncle Wilbur.

The average fraud involving a lost or stolen credit card is in excess of $1,000. Although federal law limits your loss to $50 if your card is lost or stolen, credit card companies pass their losses on to you and me in the form of higher merchandise costs, service charges, and fees. It behooves us, therefore, to promptly notify them when our card is missing. Some organizations, for a fee, will notify all your credit card companies if you lose your cards or they are stolen. You can do the same thing yourself more cheaply by keeping a list of names and numbers in a secure place.

If blank checks are stolen, call your bank immediately and then visit the bank to close your account. A thief may write checks on your account, and while the bank may be responsible for payment, you may have to prove they were forgeries.

If your keys are stolen, have your locks changed. Although it may be unlikely that the thief will show up at your house, since you don't know his motives, you should assume the worst.

There is trauma related to most crimes. If you are robbed at gunpoint or if your home is entered, whether you are in it at the time or not, there may be psychological effects upon you and others in your family. Certainly, if you are raped or injured, there will be physical trauma and a high probability of psychological trauma. In those cases, do not hesitate to seek psychological or psychiatric assistance, and be aware that others around you who have been victimized may also need such assistance and may be reluctant to seek it. Victims of sexual assault often experience sexual problems of their own as a result.

A National Institute of Mental Health study found that women who were rape or incest victims are nearly five times as likely to develop sexual problems as women who have never been assaulted. More than 50 percent of the victims studied were experiencing some sexual problem, compared to slightly more than 12 percent of the control group. Rape crisis centers provide critical assistance to victims of sexual assault; they provide victims with medical and psychological assistance and with supportive counseling from women who have experienced similar trauma themselves.

If you are a crime victim, preserve what evidence you can for the police. If your home has been broken into, make an inventory of missing items; the police will need that. But don't unnecessarily touch anything until the police have been there or have told you that they do not intend to respond. If they do respond, don't expect to have all your walls, windows, and doors dusted with black powder in order to locate fingerprints. The use of fingerprints in the solution of crime, while very important, has been overblown in television and movie drama. Fingerprints can be a vital investigative tool, but the police must have someone to match the fingerprints to if, in fact, any prints have been left at the crime scene.

In the case of rape, there are some things that you should not do. Don't bathe, douche, or wash your clothing. Go to a hospital, tell the personnel there what happened, and allow them to examine you, taking whatever samples of body fluids and tissue the police may need for evidence. While this process may sound demeaning or degrading, it may be essential to the investigation of the crime or prosecution of the criminal. For example, penetration may be a necessary element in the crime of rape in your jurisdiction. The presence of semen in the vagina supports the conclusion that penetration was made.

If you are assaulted, the police may need to photograph you without your clothes. Bruises don't show up well when covered by cloth, and they often heal before a trial begins. The photographs will have to be introduced in evidence to prove the existence of the injury. Again, cooperate in this process, as difficult or embarrassing as it might be, to aid in making sure justice prevails.

Finally, if you have been a victim, spend some time thinking about how you might have avoided it. Then do whatever you can to prevent it from occurring again. Lightning can strike twice, despite the old myth to the contrary. Then, go on about living. Life is too short to spend it kicking yourself over what you might have done differently.

13 Dangerous Strangers

The way to cure a child of fear of the dark is not to deny the existence of dark, but to walk with him in the dark and show him by example the restful quiet of it, and show him, too, how to avoid breaking his neck by stumbling over something he can't see.

Robert G. Mood, *Elementary English*

In our culture, children are prized. They are more important to us than our most valuable possessions. They are the ones whom we most cherish, and for whose safety and protection we are more concerned than for our own.

I borrowed the title of this chapter from a program I and others conducted in the elementary schools when I was a police officer. The children would assemble, and we would offer them some of the same advice, though in more basic words, that I'm offering to you. I suggest you also provide this advice to your children.

For safety's sake, children should avoid contact with people they do not know well. You must define the term ''stranger'' for your child. Parents are not strangers, some relatives are not strangers, *their* teacher is not a stranger, a uniformed police officer may not be a stranger, but the definition of who is not a stranger should be a very limited one. Since your personal situation determines who is or is not to be defined as a stranger, you must make that decision and communicate that definition to your children.

Children should be taught that they should not talk to, go anywhere with, or take anything from strangers. But, the instruction must be done in such a way that they develop a healthy suspicion, not a fear, about people they don't know well. Developing a fear of strangers will not prepare children for a happy, successful life in our society.

The discussion of strangers which ought to take place between child, parent, or teacher might begin something like this: "It's important, for your safety, for you to follow some rules that you and I are going to set. The rules are: don't talk to people you don't know; don't ever accept candy, money, or anything else from someone you don't know, unless I have told you that it is all right to take something from that person; don't let someone you do not know touch you; never go anywhere with someone you do not know; if someone you do not know tells you that your mother, father or someone else sent them to take you home, don't go with them unless you have talked to me first and I have said that you should do that; if someone tries to give you something, touch you, or take you somewhere, tell your teacher, baby sitter, a policeman, etc. about it right away."

Thorough discussion and explanation of these rules is both necessary and desirable. Our function as parents, guardians, and teachers is to prepare children to survive in this wilderness we call a society and to do that as happily as possible. Survival will be more difficult, mentally if not physically, if we instill in a child inordinate fears or "hang-ups" that follow him or her into adulthood.

We should carefully explain to our children that, while most strangers are not bad, a few are. Because we cannot tell who is and who is not bad by looking, it's best to avoid strangers altogether. Therefore, one does not stop to talk with strangers, one does not go anywhere with a stranger, one does not accept anything from a stranger, and one does not allow a stranger to touch one. I've repeated these rules purposely to reinforce them, and you should do the same with your children. Then, you must encourage them to tell you about any contact with a stranger, which means you must become a good listener, a subject we'll discuss in more detail later.

There are those who argue that the dangerous stranger approach is not an effective one, since fewer than a third of the 790,000 cases of child abuse reported annually are committed by strangers, and at least 30 percent of the time a child molester is a relative. But, my purpose here is to provide advice to protect you and your loved ones from those you don't know, not those you do know.

Your child must know where to turn for help if he or she is approached by a stranger. You can teach children to go to a police station, firehouse, school, church, or neighborhood business for help. Also explain and demonstrate the use of a pay telephone or police or fire emergency call box. Your child may be a little young for "mad" money, but should always carry, and never spend for other things, an emergency coin to activate a pay telephone.

While the home or school environment is generally a safe one, teach your child the basics about locking doors while at home, not opening doors for strangers, not letting anyone in the house without your permission, and not giving information to a stranger over the telephone. If you do not take your children to or pick them up from the baby-sitter's, nursery school, or public school, teach them the appropriate techniques we've discussed about safety while on the streets and in other public places.

If your children are going to walk to school or the baby-sitter's, you should walk the route in advance. If you find it satisfactory, walk it again with your children until you are sure they know it well. Point out safety hazards. Instruct them to walk in crosswalks and obey traffic signals when crossing streets and to walk—not run—when doing so. Don't permit shortcuts through alleys and vacant lots. Insist that the route you have walked together be followed, but point out acceptable alternatives if that route should become blocked for some reason. Point out the places children can go for help if they are in trouble.

See if there are other children who have the same destination, so that the children can walk in a group. Make sure your children know their full names, address, and telephone number; they also need to know your name and how to ask for help over a telephone if they need to do so. An identification necklace or bracelet is an excellent idea for small children, and you can include on it important medical information. Make sure your children wear the bracelet or necklace at all times, even while at home.

Impress upon your child the idea that someone in authority must know where he or she is at all times. Therefore, the child must go straight to school or another destination, and return straight home from that destination, at all times. If plans change for any reason, the child must let you or another designated person know immediately, and is not to go anywhere without advance permission. Although geography may be a factor in this decision, it's not a bad rule that the child must return home from school first, before being allowed to go to another child's home to play. That way, there is a fixed time every day when you expect your child to be at home. If he or she does not appear at that time, you know something may be wrong.

Although it may be known by another name in your community, there is a program known as the block parent program. Householders who are usually at home during the day volunteer their homes as a safe haven for children who encounter problems. Those homes are identified by clearly distinguishable window signs, but it's a good idea to have your children meet the block parents in your neighborhood *before* an emergency arises. Block parents are usually parents or grandparents themselves, are often registered with the police, and in many cases have been investigated by the police prior to becoming block parents. Don't take that for granted, though. Make sure you're comfortable with the block parents in your neighborhood, but don't be too picky. Remember, they're performing a valuable service for you and your child.

If your child does not show up within a reasonable time after school or an outing, or disappears from home, call the police, report your child missing, and ask for assistance. The police will want to know what the child was wearing when last seen, so it will be helpful for you to practice remembering how your child is dressed. After you talk with the police, stay put. Often, a lost child is quickly found, but the police may spend hours reuniting the family because Mom and Dad are scouring the neighborhood conducting their own search. So call the police, await their arrival, and then stay where they can find you.

Once you have called the police, or perhaps even before, look in all the nooks and crannies inside and outside the house to see if your child is hiding there. I have been involved in many lost child searches that lasted hours, only to find the child hiding in, under, or outside the home. The motivation for such behavior varies.

Often a child is hiding because he fears punishment for something he has done or perceives he has done, or is attempting to punish the parents for something he believes they have done. Don't be surprised if the police insist on searching your home first before looking for the child outside.

When you are with your children, make sure you watch what they are doing. Each week the daily newspapers chronicle the excruciatingly sad stories of children who have fallen into swimming pools, been backed over by a family vehicle, suffocated in an abandoned refrigerator, or disappeared because an adult whose responsibility it was to watch the child didn't know where that child was. When your child must use a public restroom, accompany him.

Never leave children alone in a car. I've investigated a number of accidents where a child left in a car released the parking brake or took the car out of parking gear and allowed it to roll, sometimes with disastrous results. Even if the car remains parked, a child is not safe in it. A vehicle can become a hot house or a refrigerator, depending upon the temperature outside. A child left inside a car can experience a rapid change in body temperature. A temperature change of just a few degrees can be fatal to a child. Also, a child left alone in a car is an easy target for kidnapping. Some kidnappings have even been accidental, occurring when a thief has stolen a car in which a child has been left, without even knowing the child was in the car.

More than a million children leave or are taken from their homes each year in this country; 90 percent of these return within two weeks. An estimated 25,000 to 100,000 are taken by divorced or separated parents, and an estimated 100,000 children disappear each year. Children are much more difficult to locate than a stolen automobile or television set. Since we don't engrave serial numbers upon our children, they're difficult to identify even if they are found. Clothing worn by kids is similar; hair styles and colors can be changed; and physical characteristics may change rapidly during a child's growing period.

People steal children for a variety of reasons, none of them honorable. Divorced and separated parents take children from their homes for reasons of love, hate, spite, or revenge. There are black market adoption rings, which sell children to those who cannot have them or cannot legally adopt them. Although precise numbers are not available, some children end up in prostitution or pornography rings. Of the missing children, a grim 2.5 percent are murdered each year.

To lessen the chance of abduction from home, law enforcement agencies recommend that a child's room be located so it is not easily accessible from the outside of the home. Although most bedroom doors should normally be kept closed to slow the spread of fire should it occur, children's rooms should be left open so that parents can hear any unusual noises.

Most missing child cases are investigated by local police who are limited by scarce resources and relatively limited jurisdictional boundaries. The Federal Bureau of Investigation becomes involved only when there is proof of kidnapping or evidence that a child has been taken across a state line.

However, legislation approved last year authorized the entry of descriptions of missing children into the FBI's National Crime Information Computer.

There are private groups which assist parents whose children have disappeared. They include: Child Find, Inc., in New Paltz, New York; Find Me, Inc., LaGrange, Georgia; National Coalition for Children's Justice, Yardley, Pennsylvania; SEARCH, Englewood Cliffs, New Jersey; Dee Scofield Awareness Program, Inc., Tampa, Florida; and Family and Friends of Missing Persons and Violent Crime Victims, Seattle, Washington. This list is not inclusive, and its inclusion here is not an endorsement of the organizations or their methods. It is an effort to offer a helpful resource to parents of missing children.

To assist in locating and identifying your children should they be taken or disappear, have photographs of them taken often, so you can provide authorities with a current picture. Keep a record, or even photos, of identifying marks such as birthmarks or scars your child may have, and keep footprints, fingerprints, and dental records of your child. Some school and police agencies will record fingerprints of youngsters for use by parents and law enforcement personnel should a child disappear.

If you are going to place your child in the care of a baby-sitter, day-care center, or nursery school, obtain references and check them before you leave your child in their care. If the individual or school is required to be licensed by the state or locality, insist on seeing proof that the license exists and is current. Establish clear rules in advance as to what your child will or will not be allowed to do under someone else's care, and what the person in whose custody you are leaving the child will be allowed to do to or with your child. Establish frequent meaningful communication with the proprietors, so that they clearly understand, by example, that you intend to be kept well informed about your child's activities. Make sure that no one but you or someone you authorize is allowed to remove your child from the nursery, school, or home.

If you leave your child in the care of someone else for any length of time, make sure you have furnished that person with proper medical releases that will authorize doctors or hospitals to treat your child if he or she should be injured or become ill. Without that, treatment may be delayed or withheld completely. If the injury or illness is critical, juvenile authorities may be compelled to make your child a ward of the court in order to provide treatment, a process that, because it consumes time, could be harmful to the health of your child.

Listen to your child. Recently in our community, the operator of a day-care facility was convicted of the death of one child and the physical abuse of a number of others over a period spanning several years. Why didn't the situation come to light sooner? Don't blame the children. Many of them told their parents about being abused by their "teacher," but those parents who listened either weren't convinced or accepted the explanations of the "teacher" without notifying authorities.

Set aside time each day to review with your children the activities of their day. Let them do the talking—you listen. If you hear something that alarms you, check it out. Don't overreact to what you hear, but remember that most small children are honest and will relate to you their sincere perception of what occurred. While children may tell tall tales, dishonesty is learned behavior. If you can't trust what your child tells you, you might spend some time in front of the mirror. While it is prudent to investigate prior to making accusations, it is imprudent to ignore what your child tells you.

"Listening" includes observing nonverbal behavior. In the case I mentioned, the mother of the child who died related that her son cried each day when he and his mother pulled up in front of the day-care home. Another little girl cried and yelled when the car in which she was riding turned into the baby-sitter's street each day. When the car stopped in front of the house, the child would stifle her tears and become stiff and nervous. She had to be led by the hand to the door. The mother of the deceased child noticed suspicious marks on her son's body several times before his death. Other children told their parents of being choked because they talked, or of being hit, kicked or shoved. Many had unusual marks on their bodies from time to time. Some parents questioned the operator of the home, who always gave a palatable explanation. But no one went to the police until a child died.

Look for obvious signs of abuse. Bruises are common among young, active children, but if they appear too frequently, there's something wrong. Be alert for behavioral patterns and changes that have no apparent explanation.

If you believe some impropriety has occurred with your child, report it immediately to your law enforcement or social services agency, or to the district or prosecuting attorney's office. They will investigate before taking any action. Cooperate fully with them, and assure them that your child will be allowed to talk with them. Help your child by being reassuring and present whenever anyone is talking with him.

One little boy asked his mother just how much he was worth. She replied that he was worth more than a million dollars to her, whereupon he asked if he could borrow a dollar of that. Most of us who are parents or grandparents, who have nieces or nephews, or who just plain love kids, would agree with her valuation and chuckle at his entrepreneurial skill. Since we value our children so highly, spending a little time to make sure they understand the rules of survival, and that we understand them, will provide a most worthwhile return.

14 Scams and Shams

A story is told about P. T. Barnum, who, while running his New York museum, became exasperated with the crowds of people who would stay in the museum for hours, leaving no room for new customers to enter. To move the spectators along and make room for more paying patrons, Barnum had a large sign painted, which he hung over an otherwise unmarked exit to the street. His scheme worked, for people were soon flocking through the exit to see the "egress."

Barnum may have been one of our better known confidence men, but he was neither the first nor the last of the breed. Barnum appealed to his patrons' sense of curiosity, and no one suffered much from his schemes. Today's confidence men and women often appeal to their victim's desire to help others and, conversely, to their greed. A con artist normally will not physically harm his victims. His weapons are a well-thought-out scheme, preparation, charm, and a convincing manner. Armed with these resources, a confidence man can rip you off for much more than a burglar or robber would get.

Let's take a look at some classic frauds. One is known as the "bank examiner" or "bank auditor" scheme. A typical version might take place like this.

A respected bank customer, often an elderly person with a sizeable savings account, is contacted by a person representing himself or herself as a bank auditor, a member of the bank's security staff, or a representative of a law enforcement agency. The "mark," as a con artist calls his victim, is told that the bank suspects one of its employees of embezzling money from the accounts of its customers. The mark is asked to cooperate with the "investigators" in order to catch the thief. Although the mark frequently indicates a willingness to assist, to sweeten the pot a reward is sometimes offered in exchange for cooperation.

The customer is asked to go to the bank and withdraw a substantial sum of money from his personal account. The phony investigators may accompany him to the bank, or arrange to meet the gullible victim afterward. They may then drive him or her home, offer profuse thanks for helping, and kindly offer to redeposit the money the victim has withdrawn. Great emphasis is placed upon the inadvisability of leaving the money in the home, where it might be stolen by thieves. Whether or not the mark is given a receipt for the deposit, in the majority of cases, neither it nor the phony bank examiners are seen again. In a few cases, the schemers are caught and the victim may see them in court, but the money is usually not recovered.

In a variation of this scheme, the mark will exchange his cash for a package handed to him by one of the investigators. The package supposedly contains an amount of cash equal to that handed over by the customer, but in marked bills which the customer is to deposit in order to help trap the embezzler. When opened at the bank, the package will contain nothing more than newspaper cut to the size of currency.

Often the motivation of the customer in this type of scheme is a sense of civic responsibility or a desire to cooperate in the apprehension of someone who is dishonest. In another well-known scheme, the "pigeon drop," the motivation of the victim is greed.

A pigeon drop involves a victim and two or more confidence persons. In a typical case, a female victim will be approached by a woman who tells the victim that she has just found a purse. Together the finder and the victim will examine the purse and find a large sum of money or a valuable piece of jewelry, but nothing to identify the owner. They will discuss at some length what should be done with the booty. Should they keep it or turn it over to the police? Unable to agree, the swindler will suggest they take their find to an acquaintance (who may be described as a lawyer or minister) and ask for advice. The advisor will suggest that the victim hold the purse and its contents for a later division among the finders, but that the victim give the other finder a sum of money as a deposit to assure that the contents are evenly divided at the agreed upon time. Once the deposit is placed in the hands of the advisor, she and her confederate are not seen again. A switch will have been made, and, the contents of the purse will prove to be worthless. There are variations on this theme, but they all involve putting up a deposit in order to share in the splitting of valuables later.

A similar scheme revolves around convincing the victim that he or she has won a large prize, or is about to receive an inheritance, but must post a cash "bond" with the swindler in order to assure delivery of the amount promised. Once the bond is posted, of course, that's the last the victim will hear of the windfall.

Those who engage in questionable business practices or illegal activities can themselves be the victims of swindlers. The "betting" or "wire room" fraud is one in which the victim, a gambler or would-be investor, is allowed to win money by making bets or investing in long-shot stock market deals with the help of tips or inside information supplied by the schemers. When the swindlers have gained the confidence of the victim by making sure his previous transactions have paid well, the victim is induced to make a substantial cash bet or cash buy. Unfortunately, something will go wrong with the deal. Phony police may raid the gaming parlor or the previously "good" stock market information may suddenly turn sour. In any event, the victim and his money will have parted company.

With the increasing popularity of illegal drugs, many would-be buyers find themselves deprived of their money in exchange for drugs that are placebos, or find their ''buy'' interrupted by the sudden presence of ''thieves'' who relieve them of their money.

Most of us have heard of the police ''stings'' in which law enforcement officers, posing as ''fences'' of stolen property, buy stolen goods from thieves. The thieves are ultimately arrested by their willing coconspirators. But the crooks aren't always the only ones who get stung. According to newspaper reports, in an effort to expose corrupt judges, the Federal Bureau of Investigation in Cleveland paid thousands of dollars in bribes to people who posed as judges, but were not. Some citizens may also have been stung by some of the con men the FBI has used to catch other crooks. The *Wall Street Journal* reported that what one con artist learned about the ABSCAM operation while working for the FBI helped him defraud dozens of people of $150,000, and another used his FBI furnished cover to sell worthless construction bonds.

There is an ageless and simple scheme known in the vernacular as the ''whoreless pimp'' scam. A young man meets a gentleman on the street who offers to furnish, for a fee, the services of a young woman. The accommodating gentleman offers to hold the man's wallet while the man visits the woman, ''for that kind of woman can't be trusted when there is money around.'' The young man is directed to a room, but if the door is opened at all, it is opened by someone who tells the young man, in language he has no trouble understanding, that he won't find what he's looking for there. The accommodating stranger will, of course, have disappeared with the wallet.

Seers and fortune-tellers who will bless your wallet to bring you good fortune are frequently found in tourist areas. But, it is usually the seer who accumulates the fortune; you may have to wait until you figure out a way to replace the money that was in the wallet before it was blessed.

Citizenship is not a requirement for being bilked in our country. Recent congressional testimony revealed the existence of a group of con artists who bilked illegal aliens of an estimated $1.5 million by selling phony Social Security cards and other documents for $20 each. Phony immigration "experts" are selling fake amnesty papers for as much as $3,000 apiece.

Confidence people are nothing if not creative. In a nearby community recently, there were more than a dozen complaints about a man who rang the doorbell and, barely holding back the tears, told the apartment dweller that he had just moved in next door and his pregnant wife had been in an accident and was at the hospital. To make matters worse, his car was in the shop and he needed to get to the hospital right away. The helpful victims usually offered to drive the man to the hospital. When they arrived, the distraught husband rushed into the emergency room, but then quickly returned to the car to say he needed money in order to admit his wife to the hospital. While the amounts he was given were small, in the majority of cases the cooperative victims handed over some cash.

In an elaborate but not so humorous prank in San Jose, California, an unknown person recruited a seventy-five man demolition crew from the state unemployment agency, then had them tear down a vacant house whose owner didn't want it torn down. The homeowner was out his investment; the workers, their pay.

Your time of grief may be a time of rejoicing for the confidence man who reads the obituaries. He may visit the next of kin to explain that just before the poor soul died, the deceased had made a down payment on an expensive gift for the surviving relative. Touched by the thoughtfulness of the departed, the grieving relative will pay the remaining balance. Of course, the merchandise was never ordered in the first place, and may not be delivered after payment is made.

If we drive, a risk we're all subjected to is the fake automobile accident. In this scheme the accident is staged by a driver who deliberately runs into your car and fakes injuries in order to collect from you or your insurance company. The fancier your car and the more prosperous you look, the more likely this person will be to pick you as a target. A variation on this theme is a group which stages the entire accident using several vehicles driven by members of the gang. They don't need your participation for this one; they'll run into each other. One of these gangs, arrested in the San Francisco area, had sixteen members and had filed 109 phony insurance claims before being arrested.

Commodities scams are those in which tricksters sell phony options to buy or sell a commodity contract and accept an "up front" premium for the option. The crooks take the premiums, never secure the contracts, and disappear. The average person has no business in the commodities market, but those who do trade in it should transact business only with established brokers.

Not all confidence schemes are complicated. A federal study estimates that the motorists of this nation spend $50 billion annually for auto repairs, and that half that amount goes for unneeded or fraudulent work. Misrepresentations are made with regard to the necessity for the repairs in the first place as well as with regard to the quality of the work performed compared to what was promised. Sometimes, no work is done at all, but the customer is billed for the work that was proposed.

Work-at-home schemes defraud several million people each year. Promoters offer large profits, indicate that no experience is necessary, and forecast large demands for services like addressing envelopes, raising worms, and clipping newspapers. The cost of the information or materials the promoter sells to his victims often exceeds the profit potential for the plan.

There are also charity scams. We have been taught that we have a responsibility to help fellow humans who are less fortunate than ourselves, and we sometimes become willing suckers for those who play upon our charitable instincts. Take the friendly Bible salesman who will solicit your contribution in order to be able to distribute free Bibles to the lost souls of the world. In one such scheme uncovered by authorities, none of the money collected went to provide Bibles. In another, the company claimed that a whopping two cents of every dollar collected went to distribute the free Bibles, but that overwhelming percentage could not be verified.

Many of us have answered our door to find a young boy or girl selling candy bars to finance a trip to summer camp, the building of a new neighborhood center, or some other worthwhile cause. In some (but not all) of these cases, there is no worthy cause other than the nearby adult who pockets the profits of the sale and who may or may not share them with the young salesperson.

The selling of tickets to circuses, sporting events, or other attractions so that needy children can attend the performance at no cost often benefits the promoters more than the would-be recipients. Sometimes there is no performance at all. In other cases, a very large percentage of the proceeds goes to the salesmen and promoters.

There are ways to protect yourself against these scams and still give to charitable organizations. First, don't give cash to people you don't know. Mail your donation in the form of a check to the sponsoring organization, and make your check payable to the organization, not the solicitor. Check with the Better Business Bureau or your consumer protection agency if you have any doubt about the legitimacy of the organization or the solicitor.

When I receive a telephone solicitation from someone claiming to be a representative of a charity, I always ask the caller to send me an explanatory letter or other literature. They seldom do, and I usually don't hear from them again. That kind of solicitor prefers to get my commitment to give, and likes to stop by to pick up my donation or have me mail it with no questions asked. I prefer to give to reputable charities which don't mind delaying a few days while I make sure they are what they say they are.

We're all fighting inflation—an economic condition for which we dearly pay. But there are those who would have us pay twice, or more. For the schemers of the world, inflation is an opportunity. High interest rates and increasing home values have precluded many potential home buyers from purchasing property. Enter the con men and women who promise potential buyers, sellers, or investors low interest loans in exchange for advance fees. The loans never materialize. The fees are not refunded. The "brokers" disappear. Some schemes are more elaborate than others; tricksters have gone so far as to set up phony escrow or trust accounts to persuade their victims that the deal is legitimate.

To fight back, make sure you deal with local, well-established escrow companies. Don't deal in cash. Reputable loan brokers are normally paid from loan proceeds at the close of escrow. Make sure the source of any low interest loan is disclosed to you in advance. If you don't know the people with whom you are dealing, check their references and ask financial people you know and trust (bankers, accountants) about them. Check with the Better Business Bureau or a consumer protection agency.

Since money isn't worth as much these days as it once was, someone may convince you that you should be dealing in gold, silver, or gemstones. They will offer you prices that are too good to believe, and you shouldn't. If you receive anything for your money in these deals, it may not be gold, silver, or diamonds in the grade or quantity you expected or paid for. You may get counterfeit coins, bars, or chips of precious metals or synthetic gemstones.

The technique of substituting fake precious metals for real is as old as the precious metals themselves. In 1872 Mark Twain wrote of the "salting" of gold and silver mines in Virginia City, Nevada Territory. One outrageous example was the North Ophir mine, which yielded "black, bullet-looking pellets of unimpeachable 'native' silver." On one of the lumps of silver could be discerned a portion of the legend "United States of," revealing that the mine had been salted with melted-down half dollars in order to increase the value of its stock.

Some deals are arranged, not to sell you something, but as a ruse to rob you of the cash you brought along with you.

Transact business only with reputable dealers who have established business locations. Don't make deals in private homes, hotel rooms, or on street corners. Have someone you know and trust verify the authenticity of the items you are buying before you pay for and accept them. Be wary of a switch—a deal where the real goods are replaced with counterfeit goods just before you go out the door.

Then there are plain old deceptive business practices which, according to the Justice Department, bilk consumers of billions of dollars a year. These include schemes which can cost corporations millions, "bait and switch" fraud at the local store, and the sale of inferior goods and shoddy workmanship in repair jobs.

Shady businessmen who give us the business are often very creative. Spurious firms collect money for the publication of manuscripts written by fledgling authors, but the texts never appear in print. An eastern photographer advertised for models and then charged them for the photographs he took of them. Unemployed people have been sold lists of job opportunities that were nothing more than a compilation of old help-wanted ads from newspaper columns.

For a long time, real estate has been a relatively secure investment which has appreciated in value, so it's no wonder that schemers are abundant in the real estate field. In addition to the more traditional mail-order sales of nonexistent land or lots in the middle of a desert or a swamp, a relatively new type of scheme is the sale of fractional interests in vacation condominiums that do not exist. Deluded into thinking they are buying an interest in real property, the disenchanted victims find that they own memberships in a vacation club.

For those of us looking for a place to live at an affordable price, there are apartment rental scams. For example, the "owner" of an apartment advertises it for rent at an eye-catchingly low rate, but doesn't advertise the address. He'll arrange to meet with you and will show you the outside of the apartment, since he doesn't want to disturb the present tenants, who are really nasty people. You won't know there's anything wrong until after you've paid the deposit and the first month's rent in cash and find that the key you were given doesn't work, or that you are one of a half dozen people who are trying to move into the apartment on the same day.

For those looking for credit but unable to get it because they have a poor or nonexistent credit record, there is the firm that promises to obtain bank credit cards for its clients in exchange for a fee and without the usual credit check. They deliver, but the catch in this case is the requirement that the customer place on deposit in the bank that issues the card an amount equal to the credit being extended.

Businesses are not immune from being fleeced. The bankruptcy scheme involves con artists who buy a small, established business and then place large orders for merchandise from suppliers who have done business with the company for years. The new owners quickly unload the merchandise, ignore the bills, and file for bankruptcy long before the suppliers catch on.

Confidence men offering large, low interest loans in exchange for a finder's fee first contact a well-known bank and ask a banker to serve as transfer agent for a large sum of overseas money. The banker happily furnishes a letter agreeing to act in that capacity. The confidence man shows the letter to his prospective victim to assure him the money will be available as promised. Once the finder's fee is paid, the con man skips town. Businesses are not the only victims of these schemes; the governments of some small countries have also fallen prey to the low-interest money finders.

Our own government is not immune from fraud, either. The Pentagon estimates that our military loses a billion dollars a year in schemes ranging from paying for $34,000 worth of pizza that was never delivered to the scrap dealer who somehow had $2 million worth of military property that didn't belong to him.

I recently read of the swindling of an Australian businessman who had invested in a nonexistent Mexican ranch where cats were to be crossed with snakes. The cats were to be bred with snakes that shed their skins twice a year, and the resulting offspring were supposed to shed their fur semiannually. Presumably, there was a market for the cat fur.

There are dozens of similar schemes, many simply new versions of old scams. They add up to lots of opportunity to lose your money to a smooth talker. So, how should you protect yourself?

The Justice Department suggests that new laws are necessary. I suggest there are more than enough laws on our books now. We're not getting ripped off by illegality, but by our own gullibility or stupidity. Once again I suggest we help ourselves, and here's how.

You will not be targeted for most schemes unless the con man believes that you have money. Keep your business dealings to yourself, don't let others know what your financial resources are, and, if you use advisors, use those you have personally investigated. You will be less likely to be approached by a bilker who is looking for a big deal.

Remember, and you can use whatever cliché you like, that there's no such thing as a free lunch or a free anything else. If a deal sounds too good to be true, it probably is. Few people are willing to give you something for nothing. Even the "free" offers you get from local businesses or major manufacturers have strings attached. They are made to entice you to shop at a certain store or to buy a certain product. The giver plans on getting his money back and more.

Be very skeptical about anyone who offers you something for nothing or who wants you to withdraw money from the bank. There's not much legitimate business being done with cash these days. Stay away from deals that require it. Use a check for major dealings, but remember that that's not foolproof either. Never let anyone "hold" your money, either literally or figuratively.

Don't do business with people you don't know or aren't able to check on. Don't be afraid to do some checking. If you are contacted by someone who claims to represent your bank, call your banker and verify the authenticity of the call. If you don't have a personal banker, now is the time to get one. If you are told that you've won a prize or can expect an inheritance, verify it with the headquarters of the firm offering the prize or the lawyer handling the estate of the deceased, and don't put up a deposit! Prizes and inheritances don't require it. Don't be afraid to check with the police, the Better Business Bureau, or a consumer protection agency. That's what they're for.

To help reduce the opportunities for actors who stage phony auto accidents, make sure you report any accident to the police and your insurance company and get good identification from the other party involved.

Never buy real estate without first seeing it, reading the fine print in the contract, and knowing the agents with whom you are dealing. If the land is such a good deal, why do they have to sell it to someone three thousand miles away? Perhaps because the locals have seen it and know what it is really worth. If you're going to rent an apartment, insist upon seeing the inside of it, make sure the person to whom you make out your check is who he says he is and lives where he says he does. Close the deal in the apartment you are renting, making sure the key works before the owner leaves.

It's okay to be neighborly, and if you want to give money away, that's charity. But be sure to give to the charity of your choice, not someone else's. If you offer help, stop short of giving away money unless you know the person so well as to know the circumstances described are accurate.

Be careful of any deal where you must pay in advance or where you can't see the product you are getting. A picture may be worth a thousand words but words may be all you'll get. Don't buy by mail unless you know the manufacturer well and never agree to pay for something you did not order.

While credit and money are available from a number of sources beside banks, stick to those that are well known and can be investigated. For the average person, a bank or finance company will be the best source for a loan.

A word about doing business yourself. If you sell something—your car, for example—sell it on a cash basis and insist on a certified check or a cashier's check for the full amount. Never sell an item on a time payment plan unless you are fully prepared to forgo the payments that are outstanding. Don't assume that you can simply repossess what has been sold if the buyer doesn't come up with the money. The laws regarding repossession are tricky, and you could be charged with theft if you try to repossess something yourself. Worse yet, you could get shot trying to repossess it.

When you hire workmen to do a job, make sure they are licensed, and check their references before allowing them to start work. In some states, a workman can file a lien against your property if the contractor for whom he has worked on your property has not paid him, even though you have paid the contractor. It makes sense to check out your contractor thoroughly in advance.

The world is full of confidence people, crooks, and schemers, but the majority of the folks we deal with are honest and hardworking. Like us, they are trying to make a living. So, while you should conduct yourself in a prudent, businesslike manner when dealing with others, it is a mistake to assume everyone is a crook. By being prudent, you'll help to keep those who are honest, honest.

15 Where to Turn

You are not alone in the fight against crime. Not by a long shot! In addition to the thousands of people who make their livings working in the criminal justice system, there are thousands more whose business or vocation is directly affected by crime and who are willing to work to see crime reduced. Then there are the many volunteers who participate in neighborhood, community, state, and national crime prevention programs.

If you want to team up with these people, how should you go about it? Let's start at the local level. Most police agencies have crime prevention officers. Larger departments usually have a crime prevention unit (which may be called community relations, public relations, or public information), while smaller agencies may assign the responsibility to a single officer. A crime prevention officer will know others in your community who are active in crime prevention. He or she will know the participants in neighborhood watch programs, block parent programs, and similar efforts, and will know when and where meetings are held. The crime prevention officer will also be a source of resource materials such as literature and films. Public speaking on crime prevention and appearances in school classrooms and assemblies are a routine part of a crime prevention officer's job.

Your insurance agent or broker can be a conduit between you and the large insurance companies, which spend hundreds of thousands of dollars each year on preventive programs and materials. An example is Commercial Union Insurance, headquartered in Boston, which has developed a crime awareness program. Several years ago, the company began development of a major public relations campaign called "Unlocking America." It may have had a selfish interest, of course: to reduce insurance claims by reducing crime. But we all have a selfish interest in crime reduction.

The program began with an advertising campaign. Utilizing magazine ads, crime prevention literature, and film, the goal was to educate the public about the costs in both dollars and human suffering of crime. Commercial Union designed a crime prevention learning kit for use in elementary and junior high school classrooms and distributed the kits to schools across our nation and in other countries.

Then the company found a typical American community that could be used as a practical laboratory for its crime prevention efforts. They selected Texarkana, the city which straddles the border of Texas and Arkansas. They furnished money and other resources to be used by local residents to develop anticrime programs. A campaign called Awareness of Crime in Texarkana (ACT) was organized. ACT established a number of programs, including a neighborhood watch, youth and school projects, and media activities, and has been especially successful in developing cooperation between citizens, business, and government. As a fourth step, the company published a two-part manual that explains how to set up a community crime-prevention program. And last year, Commercial Union held the first of its annual national conferences on crime.

This effort is just a sample of the many programs which exist in this country. You may locate similar resources by contacting individual insurance agencies or companies, or the Insurance Information Institute, an information clearing house funded by a number of insurance companies.

That's an example of how a local contact might lead you to a national program, but let's return for the moment to your own community. In addition to the local police department, sheriff's office, or state police, you also have a local district, prosecuting, or state's attorney who will be able to provide you with information, advice, and assistance. Many cities and counties also have crime prevention agencies separate from their law enforcement agencies, which employ staff people to assist in prevention efforts.

Other local agencies have a mandate directly or indirectly linked to the problem of crime. These include community mental health programs, probation and parole systems, legal services, alcoholism and drug referral agencies, crisis clinics, counseling services, and youth organizations. You'll be able to get information from the local chapter of the National Safety Council, your local American Red Cross office, or, for consumer related crimes, the Better Business Bureau. City or county licensing agencies can also be helpful in providing consumer protection information.

Local chambers of commerce often have crime deterrent committees interested in repressing crime. Large businesses in your community may help; they are often victims of crime. Local service clubs, churches, and trade organizations may also have an interest in crime prevention in your community, and the local press, radio, and television stations are usually interested in helping.

At the state level, most attorney general's offices have a crime prevention or consumer protection unit to assist you. They are particularly active in an election year. In most states there is a state police or highway patrol unit interested in crime reduction. Some states have several consumer protection agencies, each dealing with a specific, regulated industry, that may be of help in consumer issues.

State-level organizations of law enforcement officers, district attorneys, juvenile officers, narcotics officers, and even crime prevention officers may have something to offer you. There are many state-level trade organizations that, like their national counterparts, would like to see something done about crime. They range from bankers' associations to those representing plumbing and mechanical contractors. A group like the restaurant owners may focus on crime specific to their industry, while building owners and managers may have a more general interest in crime reduction.

At the national level are a number of agencies and organizations involved in crime prevention activities. The National Council on Crime and Delinquency, with chapters in many major cities, is one. The National Crime Prevention Institute in Louisville, Kentucky, provides continuing education in crime prevention. Despite a de-emphasis in crime prevention at the federal level, various federal departments and agencies have an interest in crime reduction and publish crime prevention materials. For example, the Commerce Department is concerned about crimes against businesses, and the Transportation Department is concerned about crimes against those who transport the nation's goods.

The Justice Department, the Federal Bureau of Investigation, the Treasury Department, the Secret Service, the Internal Revenue Service, the Alcohol, Tobacco, and Firearms Bureau, and the Drug Enforcement Administration are obvious resources in the fight against crime. The Department of Justice has its National Criminal Justice Reference Service. You can subscribe free (you're paying for it already) to a bimonthly guide to new publications in the criminal justice and crime prevention field.

Then there are national trade associations. The U.S. Chamber of Commerce has published materials on white-collar and organized crime. The American Banker's Association and the Bank Administration Institute publish material on bank crimes. The National Burglar and Fire Alarm Association has a vested interest in crime prevention activities. And there are many, many others.

Organizations like the International Association of Chiefs of Police, to which most police chiefs in this country belong, and the American Society for Industrial Security (ASIS), the premier international private security organization, are logically interested in crime prevention. ASIS and similar organizations have local chapters that will provide speakers, expertise, and resource material. In the past, the National Governor's Conference has taken an active role, as has the National Sheriff's Association, the National Association of Attorneys General, and similar organizations.

The more successful major efforts are usually the result of a coalition of organizations. The WeTiP[1] program, for example, offers individual and business memberships, acknowledges contributions from more than thirty-one different types of organizations, and operates in cooperation with more than twenty-five different agencies. The Take a Bite Out of Crime[2] program is a coalition of thirty-four national organizations, including three federal agencies, two labor groups, and a number of service organizations and trade associations.

I toyed with the idea of providing you with a list of organizations which could serve as resources to you, but there are several problems in doing this. The list would be so long that it would be unwieldy. Then, because organizations are being formed and disbanded daily, the long-term accuracy of such a list would be doubtful. I'd also, without a doubt, leave some organization out. So instead, I have what I think is a better solution: Use the telephone directory. It's updated every year. Look at the list of governmental agencies in the white pages, and then turn to the yellow section. You'll find associations, business and trade organizations, chambers of commerce, human services and social service and welfare organizations, and others, listed under these categories. With a little creative thinking on your part, and a phone call or a visit, you can easily enlist the aid of others in your personal fight against crime.

1. Copyright 1981 WeTiP, Inc.
2. Copyright 1979 The Advertising Council, Inc.

16 Paper and Plastic

I predict, and I am not alone in my prediction, that we will someday be a nearly cashless society. We will pay our bills and make our transactions through computer terminals or devices attached to our telephones or television sets. It would be comforting to believe that such a technological achievement would put today's criminals out of work, but our efforts toward a cashless society haven't hindered them much so far. Checks and credit cards have merely produced new types of criminals whose specialty is paper and plastic fraud.

Bad checks account for a substantial portion of all crime-related losses to business, and only a small amount of the dollars lost through bad checks is ever recovered. Bad checks can generally be broken into three categories: stolen and forged checks; fictitious checks; and insufficient funds checks. A stolen or forged check is one which the person who attempts to cash it is not the owner of the check and is not authorized to sign on the account.

Fictitious or counterfeit checks are usually printed or prepared to resemble valid checks belonging to someone else. They most often resemble payroll checks of a major corporation. An insufficient funds check is the most common type of bad check and represents the most frequent type of bad check loss. With this type of check, the signature is that of the account holder, but there is not enough money in the bank to cover the check.

If you take comfort in the fact that businesses and banks, rather than individuals, suffer most of the bad check losses, think again. What businesses pay for is passed on to you and me in the form of higher costs for goods and services, and, because of the frequency with which people cash bad checks, it has become harder and harder for an honest individual to cash one. Besides, businesses are not the only victims. You and I can be swindled anytime we lose one of our own checks or accept one from somebody else.

Millions of people who receive federal government checks are potential victims of the mailbox thief. Government employees, social security recipients, and those fortunate enough to receive a tax refund, are handy fodder for the forgers who take government checks from mailboxes, endorse them, and cash them. Thieves like government checks, since they often take months to make their way back to the person who cashed them. After that time, who's going to remember the appearance or identity of the forger? In the meantime, you don't have the check. You may be able to obtain a replacement, but not without lots of red tape and a long delay.

Most people treat their own blank checks as worthless paper. Until a check is signed by the account holder, it's worthless, right? Wrong. When you discard blank checks on an account you closed or carelessly leave blank checks on your present account lying around, you provide raw material for forgers. The loss will not be discovered in the case of checks on a closed account until the merchant who accepted the checks, imprinted with your name, address, and telephone number, receives them back from his bank. Although he or the bank may have to stand the dollar loss, your good will with the merchant and the bank won't improve, and you may have a lot of explaining to do before the matter is resolved.

In a case where your account is still active, you will learn of the forgeries in one of two ways. Checks you have written legitimately may ''bounce'' because the payment of the forgeries has diminished your balance; or, when you receive your statement at the end of the month, you will find checks included that you didn't write. You will need to contact the bank, and the bank will begin an investigation. How fast you get credit for the forged checks will depend upon how good the forgeries are, your credibility, and the speed of the bank's investigation.

Banks no longer compare signatures on checks except those accepted at the teller window, and they don't always examine those. If a forgery is accepted by a merchant, it will first be sent to his bank, then wind its way to your bank, where it will be processed by machines which can read your account number but not your signature. Not until you blow the whistle will anyone know that something is amiss, and if you're one of those people who balances their checkbook once a year whether it needs it or not, you may be in for some rough sledding when you try to convince the bank that a check written months ago wasn't written by you.

What should you do when you're on the receiving end of a check? When and what kind of check should you accept? My wife and I are flea market and garage sale aficionados; we have visited hundreds of such sales and have sold merchandise at scores of them. We've taken checks many times, and never once has one bounced. I admit we've been lucky, but we've also always insisted on good identification from anyone from whom we have accepted a check.

''Know your endorser'' is reliable advice for the mom-and-pop neighborhood store, but somewhat impractical in our world of supermarkets and discount department stores. But there are some other rules to follow.

First, I would never accept a check for an amount I could not afford to lose. Unless I know the issuer (maker, in bank parlance) very well, that means I won't take a check for more than $50. If you were to sell a car, for example, for $3,000, you should insist on receiving either a check that has been certified by a bank or a cashiers check drawn on a bank. But, remember that such checks can also be forged or counterfeited, so you should consult with the bank to make certain they have certified or issued the check. If you want to accept a personal check, you can deposit it at your bank and not release the merchandise until your bank notifies you that the check has cleared and funds have been permanently credited to your account. Don't rely on calling the bank to see if a check is good. Even if the bank will tell you one way or the other, it will not guarantee to hold the funds for you. Somebody else's check may clear before the one written to you, and the funds may no longer be available when your check reaches the bank.

Second, you should obtain good identification from anyone whose check you accept. But, remember that identification may only establish that the person who shows it to you is who he says he is. It doesn't mean the check is good. If you accept a forged or stolen check, you also accept the loss. Identification may help you find the culprit, but you may or may not get your money back.

What is good identification and what isn't? A driver's license, passport, and military identification usually have photographs and vital statistics upon them and are usually considered to be acceptable forms of identification. Social security cards, temporary driver's licenses, voter registration cards, club or membership cards, and other items without identifying characteristics or a photo are not good identification, and there's absolutely no form of identification that cannot be forged or counterfeited.

If you should be so kind as to accept a postdated check—i.e. one which is dated later than the date it is given to you—what you have done in many states is to accept a promissory note. If that check is not good, it is a civil, not a criminal matter. As a practical matter, there are so many bad checks floating around that the police and prosecuting authorities are likely to show an interest only in extremely large checks or in people who pass a large number of them. So, don't count on the authorities to help you get your money back or to prosecute the offenders.

Never, never accept a second party check—one made out to someone else, who then endorses it and gives it to you. Your only recourse is to the person who gave it to you, who may accept no responsibility if the check is not good. You should also not accept checks that are not imprinted with the maker's name, address, and telephone number, and you shouldn't give cash back in exchange for a check. A person who offers you a $50 check in exchange for a $7 item at your garage sale should be politely declined. If he writes you a $7 check and it bounces, you're out only the item, which you had already decided you could live without. If you give him the item and $43 in change, you're also out a good part of the proceeds of your sale. Any check you accept should be drawn on a local bank.

Many years ago, as a brand new bank teller fresh out of high school, I fortunately rejected a transaction from an individual who made his living getting cash back from phony checks. He would open a small account at a bank, return a week later with a very large check, ask to deposit a small but respectable amount, and request the balance of the check in cash. His transaction didn't conform to the bank's policies, and I politely turned him down. Just as politely he left and went to another bank, where he was successful. That modus operandi is still widely used today, and often successfully. A variation of that scheme is the less-experienced con man or woman who attempts to make a deposit to a nonexistent account in order to receive cash back, in hopes that the teller will not check to make sure he or she has an account. If your bank doesn't leave deposit slips lying around, it's not to make life more difficult for you. It's because they don't want that kind of individual to have easy access to them. You now know why your bank may wish to verify your account when you want to make a deposit.

The same person who steals government checks from your mailbox can steal your supply of new checks and your checking account statement. In that way, he has checks, deposit slips, and your account information, including your account balance. If you don't receive your checks or your statements promptly, call your bank. A locked mailbox will help deter a thief, as will prompt removal of your mail after it arrives. Some people arrange to have their bank hold their statements rather than mail them, to avoid having the wrong person get hold of them.

It's not necessary for an accomplished check artist to steal checks. He can print them himself, with the name of anybody's company on them. So don't let the name of a Fortune 500 company on a check dissuade you from being cautious if you don't know the person who's offering the check. Cashiers checks, money orders, and almost everything else can be counterfeited.

How about plastic? Cards are harder to counterfeit than checks, but it isn't always necessary. Cards can be stolen from the mail or can be obtained for fraudulent use through fictitious applications, burglary, robbery, or other types of theft. And the thief isn't always the user. Some prostitutes and other petty criminals make a tidy sum stealing cards and selling them. The card isn't always necessary. Crooked merchants need have only your name and credit card number to charge fictitious merchandise to your account, send the charge slips to the bank, and get immediate credit. You find out about it when you get your bill.

Those in the credit card business estimate the annual loss through fraud to be a billion dollars. Since there are about 50 million credit cards in circulation, that's about $20 a year for every single one of us who has a card. In the New York metropolitan area alone, about 5,000 cards are reported stolen each month. One major bank reports that its average fraud on a lost or stolen card is $1,100, and most fraudulent charges are made in the first three days the card is missing. Some cards are sold from city to city and are sometimes used months after they are reported stolen.

In a major scheme in Los Angeles, members of a fraud ring worked as clerks in several stores. They jotted down the names and card numbers of legitimate credit card customers, then embossed the names and numbers on plain plastic cards. They then set up a fictitious business, filled out sales slips for nonexistent purchases, and received credit to their bank account from those slips. Before the frauds could be discovered, they withdrew the money—several hundred thousand dollars—from their bank account and closed shop.

Another ring in Boston regularly ransacked the trash from a major department store and obtained names and account numbers from the carbons that separate the copies in some credit card vouchers. They too counterfeited cards that were then used for fraudulent transactions.

We are protected by federal law from paying for any losses after we report a credit card missing or stolen, and the most we can be required to pay for losses that occurred before we reported a card missing is $50. That's assuming, of course, that the card was really lost or stolen. Once again, I remind you that we *all* pay for the loss from fraudulent use of cards in the form of higher credit card fees and increased cost of merchandise and service. One bank estimates that for every $100 you charge on your card, you're paying $.45 for fraudulent transactions that occurred in the past.

What can you do to contribute to a solution? First, don't have or carry more credit cards than you need or plan to use. About 30 percent of stolen credit cards are taken directly from an individual carrying them. Another 25 percent are taken from homes, cars or businesses. So, don't leave cards in the glove compartment of your car, a dresser at home, or your desk at work. Cards are most often lost or stolen when a person is traveling. You can reduce your chances of theft by not leaving them in hotel rooms. When you use a card, make sure you get it back and that the card returned to you is yours and not someone else's.

Since cards are as good as cash, protect them as you would protect cash. Use them wisely, and inventory them periodically. Keep a list of all cards, their numbers, credit limits, expiration dates, the family members or others you have authorized to use them, and the number to call if they are missing. As I commented in chapter 12, if you're too busy to do all that, there are services which will do it for you, for a fee.

If you loan your card to someone and they make more use of it than you had planned, you are the loser, not the credit card company or the merchant. You'll have to prove you didn't loan the card to Uncle Harry if you don't intend to pay the bill Harry ran up. The message here should be clear: don't loan your cards indiscreetly. Save all your credit card receipts, so you can reconcile your bill each month when you get it. Since disreputable employees can imprint more than one receipt while they have your card in their hot little hands, you could find more charges than you have receipts for. Promptly report such discrepancies to the credit card company.

Report lost or stolen cards immediately. Studies show that the majority of card holders do just that; most people report missing cards within a day of discovering their disappearance, but 19 percent take more than four days, and 4 percent delay more than a month. That delay gives the thief great autonomy in using the card undetected.

About 25 percent of stolen cards are recovered by merchants. I should caution you that, once you report a card missing, you should not try to use it if you should suddenly find it. You may find yourself in the hands of a steely-eyed merchant who views it as stolen and you as the thief.

Before you sign a credit card receipt, make sure all the blanks are filled in and the totals are correct. If there is an error, make sure the slip is destroyed as you watch, and have the salesclerk start all over again with a new one. If you use your credit card to guarantee a reservation and then have to cancel, make sure you get a cancellation number so that you can relieve yourself of liability if the hotel forgets to cancel your registration.

If you're expecting a new credit card and don't receive it, notify the credit card company promptly. Your card may already be in circulation somewhere. Never write your credit card number on the outside of a payment envelope or give it to anyone else. The number is as valuable as the card itself for telephone purchases or for use by those who set up phony businesses with which to bilk banks and card companies. Don't give the number to telephone callers claiming to be taking a survey for a credit-card company, or others with more creative stories.

If your cards have been stolen, or if you find fraudulent charges to your account, here's what to do after you notify the credit card company of the losses by telephone: First, write a letter to the card issuer, giving your name and account number, the amount of any erroneous charge, a description of the items or services purchased, when and where the charges occurred, and why you should not be held responsible for them. It's best if you send that letter in a manner that provides you with a return receipt—certified or registered mail. The credit card company must acknowledge your letter within thirty days, unless the error can be corrected sooner. Within two billing periods, or ninety days at the most, the company must either correct your account or tell you why they believe the billing is correct.

While you are disputing the bill, you do not have to pay the amount in question or any minimum payments or finance changes that apply to that amount. However, you must pay all parts of the bill that are not in dispute. While the fraudulent charges are being investigated, the company cannot take any action to collect the disputed amount from you, such as sending threatening letters or turning the bill over to a collection agency.

There's no question that it's safer to do business with checks and credit cards than with cash. That's why they're so widely used. But the use of paper and plastic requires adherence to this set of commonsense rules if you don't want to be cheated.

17 Those Wide Open Spaces

Ah, vacation. We've covered the things the typical tourist should watch for in chapter 8. But there's that group of rough and ready types who want to get closer to nature by camping out, sunburning themselves on the deck of a boat, or risking frostbite on the ski slopes. For them, as well as for the joggers and cyclists, this chapter provides some special advice.

When I was little, "camping out" was typically the only kind of vacation we could afford, and I remember that we never had to worry about crime at campsites or parks. Tents, equipment, food, and pots and pans could be left for hours, and theft, except by the "natural bandits"—bluejays, racoons, and bears—just didn't occur. That has changed. Today's park rangers have had to become policemen, and many now wear firearms for their own protection.

When campers check in at many campgrounds now, they are given brochures that contain hints on safety and security. You are well advised to read and heed the advice given, which usually covers the local conditions. Although the incidence of violent crimes like mugging, rape, and murder is still relatively low in parks, those crimes do occur, and they are increasing alarmingly. A more frequent crime is automobile burglary; groups of backpackers sometimes return to their parking area to find their cars broken into.

Your only defense is to not leave anything of value visible in your car. Either leave your valuables at home, take them with you on your outings, or lock them in the trunk of your car. If you can't lock them out of view, because you drive a station wagon or hatchback, at least try to cover them with something. It's pretty hard to lock a canvas tent, so consider storing stoves, utensils, lanterns, and other expensive camping equipment in your car trunk while you are away from your campsite.

Get acquainted with your neighboring campers. If you're broiling a steak, give the bone to the neighbor's dog—with permission, of course. That's an easy and friendly way to break the ice. Your neighbor may be willing to watch your campsite if you ask him to, and you should provide the same service to him. Make sure you meet the park ranger. Introduce yourself, and tell him or her where you are camping. Most campsites now are numbered to facilitate the collection of fees, so you will actually have an "address" inside the park.

Beaches and pool areas are also popular with thieves, who will patrol a beach or pool looking for cameras, radios, purses, and binoculars left on blankets or towels while the owners are cavorting in the water. If you're staying in a hotel or motel, beware of thieves who will look for a room key, memorize the number, and break into the room while you are sunning yourself.

Thefts of and from boats are increasing like most other types of crime. If your boat is of the type you cart around on a trailer, remember that sometimes both boats and their trailers are stolen. Locks and cables or chains can be used to secure your trailer to the car and your boat to the trailer. When you park your trailered boat, use a trailer hitch lock or remove a wheel from the trailer.

If you store your boat at home, lock it in your garage or keep it in a fenced and locked yard. Don't leave it on the street with a "for sale" sign on it. Neighbors may think the guy who is hauling it away bought it. If you live in a multiple-dwelling unit and have to leave the boat in a parking lot, consider some sort of alarm system or antitheft device, and park it in a well-lighted area.

Locks are as important on a moored boat as they are at home or on your car. Don't leave a key in the boat ignition or hide an extra one on the boat. Make sure the hinges are on the inside of hatches and doors; use deadbolt locks on doors; secure ports with inside locks; and use strong padlocks and hasps on all hatches. Moor your boat to a secure piling or mooring with a chain or heavy cable in such a way that the chain or cable cannot be lifted over or torn off the piling or mooring. Run the chain or cable around a stanchion or under a thwart (a "rower's" seat or brace that extends from one side of the boat to another). Use one-way bolts, lock nuts, and backup plates on eyebolts. If your boat has an outboard motor, secure it to the boat with clamping screw locks or special transom bolts.

Leave your inboard engine out of commission when you leave your boat for any length of time. You can remove the rotor or a spark plug or install a hidden cutoff switch. Lock removable items in your home or some other storage facility.

Choose a marina with a reputation for good security and well-lighted moorings. Meet the proprietors and other boat owners where you dock, and keep an eye on their boats as you would have them watch yours. Let marina operators or other boat owners know if you give permission to anyone to use your boat. They should know that, if you haven't told them about it in advance, there's something suspicious about the person who wants to use it.

You should not leave valuable items on a boat anymore than you should leave them in a car. Those that you must leave should be marked as you mark those in your home. Keep an inventory ashore of all valuable items on the boat, including serial numbers and name, manufacturer, model and description of the item. Don't leave title papers on the boat. Take pictures of your boat from several angles, and record the serial and identification numbers of the boat at home, with a second list on board.

You can and perhaps should install an alarm system on your

boat. It should be a combination burglar and fire alarm system. But it's of value only if you set it when you're alone on the boat and everytime you leave it. Finally, be a good neighbor and call the police or coast guard if you see anything suspicious occurring on any nearby boat.

Skiing is an increasingly popular sport, and advocates invest hundreds of dollars in ski equipment. Annual reported ski theft losses exceed $5 million, and many ski thefts go unreported. Most thefts occur at the front or rear of ski lodges. Others occur in and around restaurants or parking areas and at overnight lodging places.

Commonsense efforts can help to prevent ski thefts. Remove your skis from your vehicle when you park it, or use an appropriate locking device to secure them to your car, a tree, or other solid object. At resorts, use ski racks that are provided. While you are resting or eating lunch, leave your skis where you can see them. Consider storing your skis and poles apart from one another. The theft of one without the other is less attractive to a thief, who would probably rather steal a whole outfit from someone else. Record the serial numbers of your ski equipment and engrave an identification number on each piece of equipment. Report thefts promptly to the police, ski patrol, or other authorities.

Bicycling and jogging are other popular action sports. The cost of a bike has increased a great deal since I rode to the corner grocer for a loaf of bread for mom. When you buy a bicycle, record the serial number for future reference. When you leave your bike in the middle of a ride, secure it. A lock or chain slipped through the front wheel or sprocket does not provide full protection against theft. Wrap the chain or cable through the frame and the rear wheel and around a post or other immovable object. Use a sturdy, high quality padlock (and make sure it is fully closed) with a case-hardened chain. Cheap locks can be picked or broken, and inexpensive chains can be cut with pliers or a bolt cutter.

Lock your bicycle every time you leave it unattended. Never leave it unlocked or unattended for "just a minute." When you are not using it, put it in a locked basement, garage, or other storage area. Don't leave it in a yard or driveway to entice a thief. Register your bike with the police, and make sure your homeowner's or apartment insurance covers bicycles.

Theft is not your only hazard when riding a bike. All the rules that apply to walking in well-known, well-lighted areas apply to biking as well. A bike does not give you the same protection or quick getaway that a motor vehicle does, so use judgment about where you ride and with whom. Ride safely and defensively. You can see a car more easily than its driver can see you. Make sure your bike is in good mechanical condition, and watch for car doors opening into traffic or cars pulling into the traffic lane. Ride with traffic; watch for drain grates, soft shoulders, and other hazards; and pay particular attention to turning vehicles that might run you off the roadway.

Rules for jogging are similar to rules for swimming. It's best not to do it alone, particularly if you are female. Women joggers have been raped or killed while jogging by themselves in remote areas or on trails. So jog with a friend or a group of people, and dress appropriately, simply to attract less attention to yourself and your physical attributes. Know where to go for help in the areas in which you jog.

Whenever you are in the great outdoors, remember that communications facilities are limited, communication is thus more difficult, and emergency services take longer to respond and do not have the resources you might be used to in a big city. Recognize these limitations as you prepare for your outing. You must be more self-sufficient than you would need to be in a metropolis. That may mean taking more survival equipment with you, having your own communications devices, and having a plan for dealing with emergencies. Monitor local radio stations. They may be your early warning system for impending emergencies like floods, fires, and storms. They will also provide advice on what to do in such situations, and can be your link to civilization.

As a case in point, our family was camping in the Mt. Baker

National Forest in Washington one summer. We had just finished a scrumptious outdoor meal and were doing the dishes while listening to the local radio station on our portable radio. Although we weren't listening carefully, we thought we heard our name mentioned in one of the announcements. Although we initially shrugged that idea off as foolish, we finally decided to drive to the nearest telephone and call the station. We did so and learned that we were indeed being "paged" due to a family emergency that required our immediate response. Although breaking camp in the dark was not as much fun as setting it up in the daylight, we were thankful that we had received the message.

Camping, boating, skiing, bicycling, and similar activities are fun, and they are part of what makes our country a great one to live in. But don't let your guard down. Just because you're having fun doesn't mean that others are not stalking you as their prey.

18 Checking Your Lists

To help you summarize and assimilate the information I've assembled in these pages, I've included checklists of the major tips for self-preservation.

Remember that you and you alone hold the single most important key to whether or not you survive in our society. How you act, react, or don't act will largely determine whether or not you will be a crime victim.

HOME SECURITY CHECKLIST

- ☐ Do your know the crime rate in your neighborhood?
- ☐ Can your home be easily seen from the street?
- ☐ Is shrubbery kept trimmed below the window level?
- ☐ Have you removed trees, trellises, etc., that would facilitate access to a second story?
- ☐ Is your house number visible from the street?
- ☐ Are ladders kept inside?
- ☐ Are exterior doors made of solid wood or metal?
- ☐ Do all exterior doors have a peephole?
- ☐ If door hingepins are outside, are they nonremovable?
- ☐ Is the door between the house and the garage treated as an exterior door?

- ☐ Have you avoided door locks which can be manipulated by breaking glass or wood panels?
- ☐ Do all exterior doors have deadbolt locks?
- ☐ Are there auxiliary locks on sliding glass doors and windows?
- ☐ If you have an electric garage door opener, does it close the door securely?
- ☐ Is the garage door kept locked at all times?
- ☐ Have you secured double-hung windows?
- ☐ Have you replaced the glass in louvered windows?
- ☐ Were door locks changed when you moved in?
- ☐ Have you secured other means of entry, such as skylights, milk or coal chutes, or basement windows?
- ☐ If you have a window air conditioner, is it secured from inside?
- ☐ Are exterior doors and windows always locked?
- ☐ Is your mailbox locked?
- ☐ Are your outside power and fuse boxes locked?
- ☐ Do you avoid leaving notes on your door for children, neighbors, and delivery people?
- ☐ Do you arrange to have lights, radios, and so on turned on and off when you are away for extended periods?
- ☐ Do you avoid hiding keys outside?
- ☐ Do you keep valuables in a bank safe-deposit box?
- ☐ Do you restrict keys to those who need them?
- ☐ Does someone always know how to contact you in an emergency?
- ☐ Do you have someone look after your home when you will be gone for an extended period?
- ☐ Have you marked valuable property in some distinctive manner?
- ☐ Do you have a list of serial numbers and other identifying characteristics of valuable property?
- ☐ If your property is fenced, are there adequate locks on all gates?
- ☐ If you have a swimming pool, is there a lock on the pool gate?
- ☐ Are fence gates solid and in good repair?

- ☐ If you have an alarm system, do you test it monthly?
- ☐ Does your apartment building have some sort of access control?
- ☐ Have you limited distribution of your apartment keys to those who must have them?
- ☐ Does your building have good entrance lighting?
- ☐ Is the parking area well lighted?
- ☐ If on-street parking is required, is the street well lighted?
- ☐ Are hallways and stairways well lighted?
- ☐ Are fire stairs locked from the stairwell side?

NEIGHBORHOOD SECURITY CHECKLIST

- ☐ Do you know the names, addresses, and telephone numbers of your neighbors?
- ☐ Do you know your neighbors work and vacation schedules?
- ☐ Do you have a neighborhood watch program in your community?
- ☐ Do you call the police if something suspicious occurs in your neighborhood?
- ☐ Does your neighborhood have a block parent program?
- ☐ Are you involved in crime prevention in your neighborhood?

FIRE PROTECTION CHECKLIST

- ☐ Can you get out of your home quickly if there is a fire?
- ☐ Is the number of the fire department posted on all telephones?
- ☐ Do you have properly located, operational smoke detectors?
- ☐ Do you test smoke detectors monthly?
- ☐ Do you change smoke detector batteries every six months?
- ☐ Do you have an adequate number of fire extinguishers?
- ☐ Are fire extinguishers the right type for use on all fires?
- ☐ Are fire extinguishers properly located?
- ☐ Are all family members trained in the use of fire extinguishers?
- ☐ Are fire extinguishers regularly inspected and recharged?
- ☐ Do you keep a hose available for firefighting?
- ☐ Do you have a fire escape plan?

- ☐ Does everyone know where to meet outside after escaping from your home?
- ☐ Do you practice your fire escape plan?
- ☐ Are all flammable liquids stored away from your house?
- ☐ Have you removed all trash or storage that might be a fire hazard?
- ☐ If you have a fireplace, is it properly screened?
- ☐ Does your fireplace chimney have a spark arrestor?

VEHICLE SECURITY CHECKLIST

- ☐ Is your vehicle in good running order?
- ☐ Do you keep your gas tank at least half full?
- ☐ Are you completely familiar with the operation of your vehicle?
- ☐ Are you a defensive driver?
- ☐ When driving, do you keep an adequate maneuvering distance between you and other cars?
- ☐ While driving, are you alert to activity alongside and behind, as well as ahead of you?
- ☐ Do you keep your car doors locked at all times?
- ☐ Do you keep the windows rolled up whenever possible?
- ☐ Do you leave your convertible top up at night and when driving in crowded areas?
- ☐ Do you avoid getting out of your car to aid disabled motorists and others who may have flagged you down?
- ☐ Do you avoid picking up hitchhikers?
- ☐ Do you park as close to your destination as possible, in a well-lighted area?
- ☐ Do you lock your car at all times when it is parked?
- ☐ Do you avoid leaving valuables inside your car?
- ☐ Do you lock your car even in your locked garage?
- ☐ Do you always inspect your car before entering for evidence of entry or someone in the back seat?
- ☐ Do you always have your keys ready before approaching your vehicle?

- ☐ When parking in an attended lot or garage, do you take all your keys except the ignition key with you?
- ☐ Do you avoid using vanity license plates?
- ☐ Do you keep your hood, trunk, and gas cap locked?
- ☐ Is your vehicle stocked with a flashlight, flares, first aid kit, jack, spare tire, lug wrench, maps, tow chain, and other emergency equipment?
- ☐ Do you plan your routes carefully in advance?
- ☐ Do you know how to take evasive action if you are being followed?
- ☐ Have you modified your car-door lock posts to make it more difficult for someone to enter your car?

TRAVEL SECURITY CHECKLIST

- ☐ Do you carefully plan your trip in advance?
- ☐ Do you secure your baggage promptly before and after your trip?
- ☐ Do you avoid advertising that you are a tourist?
- ☐ Do you stay in well-known, reputable hotels?
- ☐ Does someone always know where you are or where you are supposed to be?
- ☐ Do you limit the amount of cash and credit cards you carry?
- ☐ Do you avoid strangers?
- ☐ Do you know how to get out of your hotel safely in an emergency?
- ☐ Do you make sure your hotel room is secure?
- ☐ Do you carry a flashlight and spare prescriptions when you travel?
- ☐ Do you make sure cab drivers do not deviate from the route you request?
- ☐ Do you avoid traveling during night hours?
- ☐ Do you avoid remote areas of unknown territory?
- ☐ Do you know enough of the language of a foreign country to use the telephone or to ask for a policeman or doctor?

TELEPHONE SECURITY CHECKLIST

- ☐ Do you list only your last name and initials in the telephone directory?
- ☐ Do you avoid giving your telephone number to those who don't need it?
- ☐ Do you avoid giving information over the telephone?
- ☐ Are others in your family trained to avoid giving information over the telephone?
- ☐ Do you hang up on nuisance and obscene callers?
- ☐ Do you call the telephone company and the police when such calls persist?
- ☐ Do you have a telephone adjacent to your bed?
- ☐ Are emergency numbers posted on your telephones?

PERSONAL SECURITY CHECKLIST

- ☐ Do you observe who and what are around you at all times?
- ☐ Do you "follow your hunches"?
- ☐ When you walk, do you do so on well-traveled, well-lighted streets?
- ☐ Do you avoid walking alone at night?
- ☐ Do you walk in the middle of the sidewalk, avoiding parked cars, alleys, and doorways?
- ☐ Do you walk facing traffic?
- ☐ Do you avoid taking shortcuts?
- ☐ Do you avoid stopping to talk to strangers?
- ☐ Do you avoid trouble when you see it?
- ☐ Do you walk confidently?
- ☐ Do you dress modestly?
- ☐ Do you carry no more valuables than necessary?
- ☐ Do you carry wallet or purse safely and securely?
- ☐ Do you always have money for a bus, cab, and the telephone?
- ☐ Do you wait for public transportation in populated, well-lighted areas?
- ☐ Do you turn down rides from people you don't know well?
- ☐ When you are riding a bus, do you sit close to the driver?

- ☐ Are all family members schooled not to open the door to delivery persons or admit unexpected visitors?
- ☐ Does your family keep a low profile about their activities?
- ☐ Does your family know what to do if they suspect an intruder is in your home?
- ☐ Do you watch for suspicious persons and report them?
- ☐ If you have firearms in your home, are they properly secured?
- ☐ Have family members had firearms training?
- ☐ Are firearms kept in good condition?
- ☐ Are locks changed when keys are lost or stolen?
- ☐ Are names and addresses left off key tags?
- ☐ Does someone know where all family members are at all times?
- ☐ When you work late, do you try to work with someone else?
- ☐ On an elevator, do you stand near the control panel?
- ☐ Do you check around you before entering or leaving an elevator?
- ☐ Do you always have your keys ready when you get to your door?

SECURITY CHECKLIST FOR CHILDREN

- ☐ Are your children taught to avoid contact with people they don't know well?
- ☐ Have your children been taught to go to teachers, policemen, or other authority figures for help?
- ☐ Do your children carry change so they can use the telephone in an emergency?
- ☐ Do your children know not to open doors to strangers?
- ☐ Have you walked the route to school with your children?
- ☐ Are your children encouraged to walk in groups?
- ☐ Do your children wear identification bracelets or necklaces?
- ☐ Do your children always let you know where they are?
- ☐ Do you and your children know the block parents in your neighborhood?
- ☐ Do you have current pictures of your children?

- ☐ Do you avoid leaving your children alone in the car?
- ☐ Do you check out baby-sitters, day-care centers, nursery schools, etc., before leaving your child in their care?
- ☐ Do you listen to your children?
- ☐ Do you report suspicious occurrences to the authorities?
- ☐ Are your children's rooms located so that they are not easily accessible from outside?
- ☐ Are doors to your children's rooms kept open so that unusual noises can be heard?
- ☐ Do you avoid leaving your children unattended at home?
- ☐ Do your children know how to call the police?
- ☐ Have your children been taught to walk along heavily traveled streets and to avoid isolated areas?
- ☐ Have your children been taught to refuse rides from strangers?
- ☐ Do your children tell you immediately about anyone who approaches them?
- ☐ Have your children been taught that they are not to leave school with anyone without your approval?

Glossary

Throughout this text, I've used some terms that are fairly commonly understood by those who work in the criminal justice system, although I admit that lengthy discussions sometimes ensue even among the professionals over the precise meaning of a given term. So that you know what I mean when I use these terms, I've included some commonly accepted definitions. They are furnished, by the way, by the U.S. Department of Justice's National Criminal Justice Information and Statistical Service. I have modified them some to make them more easily understandable.

Abscond: Intentionally and unlawfully absent or conceal oneself in order to avoid a legal process.

Acquittal: A judgment of a court, based either on the verdict of a jury or a judge that the defendant is not guilty of the offense for which he or she was tried.

Adjudication: The judicial decision terminating a criminal proceeding by a judgment of conviction or acquittal, or a dismissal of the case.

Adult: A person who is within the jurisdiction of a criminal, rather than a juvenile, court because his or her age at the time of an alleged criminal act was above a statutorily specified limit (usually eighteen).

Alias: Any name used for an official purpose that is different from a person's legal name.

Appeal: A request by either the defense or the prosecution that a case be removed from a lower court to a higher court in order for a completed trial to be reviewed by the higher court.

Arraignment: The appearance of a person before a court in order that the court may inform the person of the accusations against him or her and hear the person's plea.

Arrest: Taking a person into custody by authority of law for the purpose of charging him or her with a criminal offense or for the purpose of initiating juvenile proceedings.

Arson: The intentional destruction or attempted destruction by fire or explosive of the property of another or of one's own property.

Assault: Unlawful intentional inflicting, or attempted or threatened inflicting, of injury upon another.

Assault, aggravated: Unlawful intentional causing of serious bodily injury with or without a deadly weapon, or unlawful intentional attempting or threatening of serious bodily injury or death with a deadly weapon.

Assault, simple: Unlawful intentional threatening, attempted inflicting, or inflicting of less than serious bodily injury, in the absence of a deadly weapon.

Assault with a deadly weapon: Unlawful intentional inflicting, or attempted or threatened inflicting, or injury or death with the use of a deadly weapon.

Attorney, lawyer, counsel: A person trained in the law, admitted to practice before courts in a given jurisdiction, and authorized to advise, represent, and act for other persons in legal proceedings.

Auto Theft: *See* Motor Vehicle Theft

Booking: A police administrative action officially recording an arrest and identifying the person, place, time, arresting authority, and reason for the arrest.

Burglary: Unlawful entry of a structure with or without force, with intent to commit a crime.

Case: At the level of police or prosecutorial investigation, a set of circumstances under investigation involving one or more persons; at subsequent steps in criminal proceedings, a charging document alleging the commission of one or more crimes; in juvenile or correctional proceedings, a person who is the object of agency action.

Case, court: A single charging document under the jurisdiction of a court, or a single defendant.

Charge: A formal allegation that a specific person has committed a specific offense.

Check fraud: The issuance or passing of a check, money order, or draft that is legal as a formal document, signed by the legal account holder, but with the foreknowledge that the bank or depository will refuse to honor it because of insufficient funds or closed account (*see also* Counterfeiting and Forgery).

Citation: A written order issued by a law enforcement officer directing an alledged offender to appear in a specified court at a specified time in order to answer a criminal charge.

Complaint: A formal written accusation made by any person, but often a prosecutor, and filed in a court, alleging that a specified person has committed a specific offense.

Conviction: A judgment of a court, based either on the verdict of a jury or a judge or on the guilty plea of the defendant, that the defendant is guilty of the offense for which he or she has been tried.

Count: Each separate offense, attributed to one or more persons, as listed in a complaint, information, or indictment.

Counterfeiting: The manufacture or attempted manufacture of a copy or imitation of a negotiable instrument with value set by law or convention, or the possession of such a copy without authorization, with the intent to defraud by claiming the genuineness of the copy.

Court: An agency of the judicial branch of government, authorized or established by statute or constitution, and consisting of one or more judges, which has the authority to decide upon controversies in law and disputed matters of fact brought before it.

Credit card fraud: The use or attempted use of a credit card in order to obtain goods or services with the intent to avoid payment.

Crime: An act committed or omitted in violation of a law forbidding or commanding it for which an adult can be punished, upon conviction, by incarceration and other penalties; or a corporation penalized; or for which a juvenile can be brought under the jurisdiction of a juvenile court and adjudicated a delinquent or transferred to adult court.

Crimes of violence: A term that generally includes murder, voluntary manslaughter, forcible rape, robbery, and assault.

Defendant: A person against whom a criminal proceeding is pending.

Defense Attorney: An attorney who represents the defendant in a legal proceeding.

Delinquency: Juvenile actions or conduct in violation of criminal law.

Dismissal: A decision by a judge to terminate a case without a determination of guilt or innocence.

District Attorney: *See* Prosecutor

Embezzlement: The misappropriation, misapplication, or illegal disposal of legally entrusted property with intent to defraud the legal owner or intended beneficiary.

Extortion: Unlawful obtaining or attempting to obtain the property of another by the threat of injury or harm to that person, or his or her property, or another person.

Felony: A criminal offense punishable by death or by incarceration in a state or federal prison for a period of which the lower limit is prescribed by statute in a given jurisdiction, typically one year or more.

Fine: The penalty imposed upon a convicted person by a court requiring that he or she pay a specified sum of money.

Forgery: The creation or alteration of a written or printed document, which if validly executed would constitute a record of a legally binding transaction, with the intent to defraud by affirming it to be the act of an unknowing second person.

Fraud: An element of certain offenses, consisting of deceit or intentional misrepresentation with the aim of illegally depriving a person of his or her property or legal rights.

Fugitive: A person who has concealed him- or herself or fled a given jurisdiction in order to avoid prosecution or confinement.

Hearing: A proceeding in which arguments, witnesses, or evidence are heard by a judge or administrative body.

Homicide, criminal: The causing of the death of another person without justification or excuse.

Homicide, excusable: The intentional but justifiable causing of the death of another or the unintentional causing of the death of another by accident or misadventure, without gross negligence.

Homicide, justifiable: The intentional causing of the death of another in the legal performance of an official duty or in circumstances defined by law as constituting legal justification.

Indictment: A formal written accusation made by a grand jury and filed in a court, alleging that a specified person has committed a specific offense.

Jail: A confinement facility usually administered by a local law enforcement agency, intended for adults but sometimes also containing juveniles, which holds persons pending adjudication and/or persons committed after adjudication for sentences of a year or less.

Judge: A judicial officer who has been elected or appointed to preside over a court of law, whose position has been created by statute or by constitution and whose decisions in criminal and juvenile cases may be reviewed only by a judge of a higher court.

Judgment: The statement of the decision of a court that the defendant is convicted or acquitted of the offense charged.

Jurisdiction: The territory, subject matter, or person over which lawful authority may be exercised.

Jury, grand: A body of persons who have been selected and sworn to investigate criminal activity and the conduct of public officials and to hear the evidence against an accused person to determine whether there is sufficient evidence to bring that person to trial.

Jury, trial: A statutorily defined number of persons selected according to law and sworn to determine certain matters of fact in a criminal action and to render a verdict of guilty or not guilty.

Juvenile: Generally, a person under the age of eighteen years.

Kidnapping: Unlawful transportation of a person without his or her consent or without the consent of a guardian, if a minor.

Larceny: Unlawful taking or attempted taking of property from another.

Law enforcement agency: A federal, state, or local criminal justice agency of which the principal functions are the prevention, detection, and investigation of crime and the apprehension of alleged offenders.

Law enforcement officer: An employee of a law enforcement agency who is an officer sworn to carry out law enforcement duties, or a sworn employee of a prosecutorial agency who primarily performs investigative duties.

Maim: To cripple, mutilate or disable another person.

Manslaughter: Causing the death of another by recklessness or gross negligence.

Manslaughter, vehicular: Causing the death of another by negligent operation of a motor vehicle.

Manslaughter, voluntary: Intentionally causing the death of another with reasonable provocation.

Misdemeanor: An offense usually punishable by incarceration in jail for a period of which the upper limit is prescribed by statute in a given jurisdiction, typically limited to a year or less.

Modus operandi: Method of operation.

Motor vehicle theft: Unlawful taking, or attempted taking, of a motor vehicle owned by another, with the intent to deprive him or her of it.

Mugging: *See* Robbery, Strongarm

Murder: Intentionally causing the death of another without reasonable provocation or legal justification, or causing the death of another while committing or attempting to commit another crime.

Offense: An act committed or omitted in violation of a law forbidding or commanding it.

Penalty: The punishment annexed by law or judicial decision to the commission of a particular offense, which may be death, imprisonment, fine, or loss of civil privileges.

Plea: A defendant's formal answer in court to the charges brought against him or her.

Police department: A local law enforcement agency directed by a chief of police or a commissioner.

Police officer: A local law enforcement officer employed by a police department.

Prison: A confinement facility having custodial authority over adults sentenced to confinement for more than a year.

Probation: The conditional freedom granted by a judge to an alleged offender, or adjudicated adult or juvenile, as long as the person meets certain conditions of behavior.

Prosecutor: An attorney whose official duty is to initiate and maintain criminal proceedings on behalf of the government against persons accused of committing criminal offenses.

Rape: Unlawful sexual intercourse by force or without legal or factual consent.

Rape, forcible: Sexual intercourse or attempted sexual intercourse with a person against his or her will, by force or threat of force.

Rape, statutory: Sexual intercourse with a person who has consented, in fact, but is deemed, because of age, to be legally incapable of consent.

Robbery: The unlawful taking or attempted taking of property that is in the immediate possession of another, by force or the threat of force.

Robbery, armed: The unlawful taking or attempted taking of property that is in the immediate possession of another by the use or threatened use of a deadly or dangerous weapon.

Robbery, strongarm: The unlawful taking or attempted taking of property that is in the immediate possession of another by the use or threatened use of force, without the use of a weapon.

Sentence: The penalty imposed by a court upon a convicted person, or the court decision to suspend imposition of execution of the penalty.

Sheriff: The elected or appointed chief officer of a county law enforcement agency, usually responsible for law enforcement in unincorporated areas and for the operation of the county jail.

Sheriff, deputy: A law enforcement officer employed by a county sheriff's department.

State police: A state law enforcement agency the principal functions of which include maintaining statewide police communications, aiding local police in criminal investigations, police training, guarding state property, and highway patrol.

State's Attorney: *See* Prosecutor.

Subpoena: A written order issued by a judge requiring a specified person to appear in a designated court at a specified time in order to serve as a witness in a case under the jurisdiction of that court or to bring material to that court.

Summons: A written order issued by a judge requiring a person accused of a criminal offense to appear in a designated court at a specified time to answer the charges.

Suspect: A person, adult or juvenile, considered by a criminal justice agency to be one who may have committed a specific criminal offense.

Theft: Larceny, or in some legal classifications, the group of offenses including larceny, and robbery, burglary, extortion, fraudulent offenses, hijacking, and other offenses sharing the element of larceny.

Trial: The examination of issues of fact and law in a case of controversy beginning when the jury has been selected in a jury trial, or when the first evidence is introduced in a court trial, and concluding when a verdict is reached or the case is dismissed.

Verdict: In criminal proceedings, the decision made by a jury in a jury trial or by a judge in a court trial, that a defendant is either guilty or not guilty of the offense for which he or she has been tried.

Victim: A person who has suffered death, physical or mental suffering, or loss of property as the result of an actual or attempted criminal offense committed by another person.

Warrant, arrest: A document signed by a judge that directs a law enforcement officer to arrest a person who has been accused of an offense.

Warrant, bench: A document issued by a judge directing that a person who has failed to obey an order or notice to appear be brought before the court.

Warrant, search: A document issued by a judge that directs a law enforcement officer to conduct a search for specified property or persons at a specific location, to seize the property or persons if found, and to account for the results of the search to the issuing judge.

Witness: A person who directly perceives an event or thing, or who has expert knowledge relevant to a case.

Index